#1

RY .51

SYRA

W.V.

SANF

NYJ

PHI

MIA

MINN

LARAID

HOUS

Also by Howard Blum

NONFICTION

I Pledge Allegiance . . .
The True Story of the Walkers: An American Spy Family

Wanted!
The Search for Nazis in America

FICTION

Wishful Thinking

OUT THERE

The Government's Secret Quest for EXTRATERRESTRIALS

HOWARD BLUM

Simon and Schuster

New York London Toronto Sydney Tokyo Singapore

SIMON AND SCHUSTER
Simon and Schuster Building
Rockefeller Center
1230 Avenue of the Americas
New York, New York 10020

Designed by Liney Li
Manufactured in the United States of America

1 3 5 7 9 10 8 6 4 2

Library of Congress Cataloging-in-Publication Data
Blum, Howard, date–
Out there : the government's secret quest for extraterrestrials / Howard Blum.
p. cm.
"A leather-bound signed first edition of this book has been published by The Eastern Press"—T.p. verso.
Includes bibliographical references.
1. Life on other planets—Research—United. States.
2. Unidentified flying objects. 3. n-us. I. Title.
QB54.B64 1990
001.9'42'072073—dc20 90-38559
CIP

ISBN 0-671-66260-0

A leather-bound signed first edition of this book
has been published by The Easton Press.

For Jenny, with love.
And for Fred Hills and Burton Beals, wise friends.

CONTENTS.

A Note to the Reader

This is a true story. I verified every name, incident, date, and conversation that is recorded in this account. To accomplish this, I have relied on extensive interviews with civilian and military officials, both past and present; members of the intelligence community; the FBI; and scientists working both in and out of the government. I have also consulted confidential military documents, classified research reports, and reports prepared exclusively for the UFO Working Group, as well as research papers and government publications available to the public. A more detailed explanation of my method of attribution is included in A Note on Sources.

H. B.

The ultimate object of magic in all ages
was and is to obtain control of the sources of life.

—*W. B. Yeats*

Prologue
A SPY'S STORY

The spy refused to answer any more questions. It was nearly midnight and we were sitting in a hotel room that looked straight across Pennsylvania Avenue toward a floodlit White House. There already was a row of empty miniature scotch bottles on the coffee table in front of us, but I got up and walked across the room to the minibar.

"One for the road," I suggested. I didn't want him to leave and I figured this was a better strategy than arguing. After all, he seemed to have quite a thirst.

He accepted the drink, stirring it with a long, thin finger; and, I noted, he made no move to rise from the sofa.

"I've said all I'm going to say," he insisted.

"Sure." Some people you don't push; you just give them time.

"I've already told you plenty. Right?"

"Right." Which was the truth. For the past two hours he had been the perfect source: a man with a mission. He was a senior official at the National Security Agency and he had contacted me after reading a *New York Times Magazine* piece I had written about the John Walker spy case. The biographical squib in the magazine had explained that I was writing a book on this family of spies and the NSA intelligence officer had made the effort to track me down in New York City because, he explained, he wanted to make sure my book was accurate. Mostly, though, he was angry. Even over the long-distance phone his patriotic dismay was an open wound: "Walker told us that stealing secrets from this country is as easy as shoplifting from K mart. And I'll be damned if the bastard isn't right."

"We should talk about this," I quickly suggested.

He agreed. "The nation's got to wake up before it's too late." And he added, "Next time you're down in Washington, let's get together for a drink."

I caught the shuttle that afternoon. My plane landed in the last hour of daylight on a warm spring day in 1987, and it was dark by the time my cab made its way into the city and I had checked into my room. I waited impatiently until precisely nine o'clock and then, as arranged, went downstairs to meet him in the hotel bar. He spotted me first, and gestured with a small, very discreet nod of his head. Foolishly, I bounded across the crowded room with what I hoped was an ingratiating grin and offered my hand. He took it, but the public display seemed to unsettle him. "It's a zoo in here," he announced before I could find a seat. "Let's go for a walk."

We walked toward the Old Executive Office Building. The night was warm, yet the downtown streets were nearly empty. We talked about the cost of living in New York City, about working at *The New York Times,* about the Yankees and their owner George Steinbrenner. About anything but the Walker spy case. I let him set the pace; the source always makes all the rules. Finally, we were back at the hotel. When I walked to the elevator, he simply followed without comment.

For the next two hours, he sat on the green damask sofa in my room sipping glasses of scotch while I hurled questions. From the first, he was a ready accomplice. He was eager to share what he knew because, he said with conviction, "the public has a right to know." A moment later, though, he broke into a rare grin and noted, "Besides, if Walker knew it, it's not secret anymore." But by midnight he was no longer cooperating. As happens in most interviews—even the most collaborative of sessions—we had reached the wall: the point he had resolved not to go beyond no matter what.

Yet watching him nurse that last drink, I realized I wasn't the only one who was reluctant to call it a night. Now that he had broken a lifetime of training and, contrary to all his professional instincts, had shared classified information with A Member of the Press, he was not ready to slink off. Playing the deep background source is an intoxicating, aggrandizing role; and, as I have so often observed, it is a repetitious calling. It was in this high-spirited mood, helpful and confiding, hoping to prolong his moment, that he reached for something else to pique my interest.

"You know," he threw out, "there's been a lot of talk around the NSA about outer space. Weird stuff. UFOs."

"Uh-huh," I said evenly. I remained determined to steer him back to Walker, but for the moment I restrained myself. Rushing, I knew, wouldn't help things at all.

"Yeah," he continued. "Heard they got some kind of all-star working group or something. A panel of hotshots zeroing in on UFOs. Going to get the truth at last."

"Think there might be something to get?" I was being polite.

He shrugged massively. "I don't know," he said after a bit of thought. "But the story's going around that a lot of strange things are happening. We're catching a lot of crazy signals on our microphones and they're not from this planet. That's a fact. You might want to look into it."

And that was how it all began. That was how I first learned about the government's covert search for extraterrestrial intelligence. That was how I received the first clue that would lead me on a two-year chase across the entire country, from military bases under snowcapped mountains to installations hidden away in the midst of vast deserts, from laboratories guarded by military policemen to classified government engineering projects staffed by long-haired graduate students, from the New Mexico hill country to the woods of northern Wisconsin. That was why I first began to wonder what was out there—and what the government really knew about it.

Yet at the time I paid little attention; it wasn't the story I had rushed to Washington to hear. Instead, I affably told my source that maybe all this "weird stuff" was, indeed, a lead worth pursuing. Then, realizing it was now or never, I made one last attempt to bring the conversation back to John Walker. And this time around, much to my delight, I managed without too much effort to batter a large hole through my source's wall of resolve.

The next day I flew back to New York and sat down at my desk to write about the Walker spy family. I didn't give the NSA man's talk, an aside really, about "weird stuff" and "a panel of hotshots zeroing in on UFOs" much attention. I had a book to write. But by the time the final pages on the final diskette had spewed forth from my printer, his words—a challenge!—were resounding unexpectedly through my consciousness. Was the government back in the UFO business? Had they found anything? Was there life in the universe?

No doubt about it, I was curious. Very curious.

Notebook in hand, I began to poke around official Washington. I made a hundred calls, sat in suburban Virginia living rooms on couches covered in meadows of chintz, took notes in Beltway offices where the doors closed ominously behind me at the touch of unseen buttons, met with crew-cut, square-jawed military officials in surveillance-proof rooms, shared lunches with avuncular scientists in crowded institutional cafeterias—and along the way I began to perceive the outlines of a startling story: there was a top-secret government "panel of hotshots." It was called the UFO Working Group.

And if I had any doubts about the veracity of my original NSA contact, they were assuaged when the government, bless it, provided me with proof of his bona fides. It was not long after the publication of my book on the Walker spy family that my publisher received a letter from an indignant lawyer. He had been retained by an NSA official who was being accused by the agency of being one of my sources. The good news, I told my publisher, was that the spiteful powers at Fort Meade had targeted the wrong man. But the even better news, I rejoiced silently, was that if the NSA was so concerned about where I was getting my information, my source must have access to the Real Thing. It wasn't fool's gold that had been waved before me. I returned with renewed enthusiasm to my prospecting in the deep mines of official Washington.

Yet the details of the story I was chasing still seemed to be beyond my grasp. Facts, always a precious commodity, were very hard to come by. I needed help. That was why, hat in hand, I decided to call on Seymour Hersh. Sy and I had both put in time at *The New York Times* and like anyone who has had the experience of sharing a newsroom with this Pulitzer Prize winner, I had quickly become aware of two things. First, he was an impeccable and relentless reporter; pity the poor soul who dared to stand between Sy and The Truth. And, a dangerously close second: He was a man, as an intimidated editor had once remarked, whose bite was truly worse than his bark. So it was not without some trepidation that I went marching into his cluttered, closet-sized office in the National Press Club.

I was a model of politeness. I explained to Sy that I was working on a story about the government's secret search for extraterrestrial life. In fact, I had reason to believe that a clandestine panel called the UFO Working Group had been formed. Could Sy, for old *Times* sake, give me a hand in getting to the bottom of this?

Sy went from zero to sixty in about a second: "Get out of here,

Blum! I got enough to do without helping you check out kooky stories. You go New Age on me or something? You think I give a damn about that sort of craziness? What do I care about UFOs?"

But before he finished his tirade, he had also reached for the phone.

Two days later I was back in New York when I received a call: "Blum? Hersh here. You're right—the government is as crazy as you are. They do have some kind of committee that's looking into all that kooky stuff."

"Can you give me a name? Someone I can talk to?"

"You think I'm gonna do your work for you? There's a story here. Isn't that enough for you? Now find it yourself." And then he hung up.

So, I went to find it.

I started, a modest but necessary first step, by making formal inquiries at all the relevant government agencies. Of course, I was not optimistic about what I would learn. I expected weary, noncommittal shrugs or, at best, succinct "no comments." I was mistaken. The denials were surprisingly tenacious and absolute.

The Department of Defense spokesman took three days to mull over my question about the UFO Working Group and then reported back: "There is no record of any committee, either official or unofficial, called the UFO Working Group. Further, since the conclusion of Project Blue Book in 1969 there has been no official or unofficial investigation, study, or involvement in the matter of UFOs by any branch of the Armed Forces or the Defense Intelligence Agency."

The CIA spokesman took only a day to reply, but he spoke with similarly earnest certitude: "On the record and off the record, the agency has no knowledge of any group by the name of the UFO Working Group. The agency has no active research into UFOs or extraterrestrial life. In fact, its only involvement with UFOs was the Robertson Panel Report which was issued in 1953."

The NSA spokesman was terse, yet a bit more cautious: "I can find no one who will confirm the existence of the UFO Working Group or an involvement by the NSA with UFO investigations."

And, I soon learned, all three statements were lies.

Part I
THE UFO WORKING GROUP

ONE.

There are 7,087 man-made objects in space. They include ten screws, each an eighth of an inch in diameter, that were discarded during a 1984 shuttle mission; a thermal glove that floated out of Gemini 4 in 1965; a screwdriver dropped by a careless space-walking cosmonaut aboard the Soviet Mir space station; and several thousand satellites, some actively communicating with ground stations, and some relics, useless twentieth-century antiques that will continue to whiz and clank through space for generations. All this hectic traffic above the planet Earth is constantly monitored by the U.S. Space Command's Space Surveillance Center hidden inside snowcapped Cheyenne Mountain near Colorado Springs. And it was here, a half-mile below ground in a narrow chamber known as Box Nine, on an icy December evening in the weeks before Christmas 1986, that the first event which led to the formation of the UFO Working Group began to play itself out. For on a computer screen in that windowless room the 7,088th object in space was discovered.

Commander Sheila Mondran, a crew chief at the U.S. Space Surveillance Center, was late for work. She had spent the afternoon Christmas shopping in Colorado Springs and, much to her delight, had worked pretty far down her list. Then she noticed the heavy gray snow clouds moving in. It was after five, so the commander, without too much internal debate, decided she had better hightail it out of the mall. She was scheduled for the middle shift, 6:00 P.M. to 2:00 A.M., and the road back up the mountain could get a little tricky with a coat of fresh snow.

It was worse than she had expected. Her car kept skidding on the wet snow. And all the while the storm kept building, white streaking down on top of white. Her windshield wipers were nearly useless; distances were obliterated. Even Cheyenne Mountain seemed daunted by the storm, its familiar looming peak now only a small, vague presence.

Yet as she got closer to the Space Surveillance Center, near enough so that she began to roll down her window to let the guard at the main gate get a look at her commander's stripes, she suddenly realized how startlingly beautiful the mountain had become on this snowy night. Crystal Palace was the playful nickname the locals, a bit dazzled by all the real and imagined "Star Trek" gadgetry rumored to be housed in the subterranean headquarters, called the facility; and for the first time the commander thought the joke was particularly apt: the spotlighted mountain seemed to shine and sparkle with this fresh, polishing dust of snow. If only it hadn't been six-fifteen.

She parked her car and, instantly regretting she hadn't worn her snow boots, hurried to the military shuttle bus. She took the seat nearest the door, and the bus began to lumber down the two-lane roadway. It was pitch dark inside the bus, but spotlights continued to illuminate the mountain as if it were the main attraction at a movie premiere. There was no point in once more checking her watch, the commander decided; she could only pray it had been another uneventful shift. The driver, intimidated by the weather, continued at an infuriating snail's pace. At last, the bus approached the entrance to the command post, and the gray metal blast doors started to swing open. This, too, seemed to take an eternity. The doors were as thick as four or five telephone poles laid back to back and they moved with some deliberateness. Finally, the bus drove inside the granite core of the mountain; and the commander looked out the rear window as the huge doors, like the outside world itself, gradually closed behind her. There was a booming thud as the doors shut tight. She was locked inside the mountain.

It was a cave of fluorescent shadows. A harsh unnatural light glowed around the clock in an attempt to keep this artificial world at a constant high noon. The light bounced off the polished steel doors and walls and, to the commander's mind at least, seemed to singe. Everyone in the cave was immediately branded, their skin rendered a telltale pasty, sallow shade.

There were also, and she was certain of this, too, mind-bending consequences from constant exposure to this man-made climate of eerily strong light and contrived air (there might just as well have been only two settings: tropical rain forest or Arctic tundra). A tour in "the cave" left one detached and listless; "apathetic" was the word she took to using when she filled out efficiency reports.

This evening, though, Commander Mondran didn't mind the blight of "cave fever" that, whatever its cause, seemed to be affecting the Surveillance Center. As she took a polished steel elevator down to the third level, descending almost 2,500 feet, she was beginning to convince herself there was no need to have worried. Despite her lateness, life in Crystal Palace had continued at its slow-motion pace without her. She had not even been missed.

This realization was further reinforced when she reported to duty in Box Nine. The crew of six orbital analysts were already seated in front of their computer screens. No one seemed to notice her enter; at least these "knob turners" didn't say a word. Instead, their faces stared straight ahead at their computer display screens. Maps were etched in a strident green light on each console; across this terrain the curving ground tracks of an object off in space were illustrated. At the press of a computer button, a specific object or a certain sector on the globe could be isolated. It was very dull work. Routine had neutralized the job. An eight-hour shift seemed to stretch forever.

The commander, within minutes of entering windowless Box Nine, was overwhelmed with weariness. She sat at the crew chief's desk and tried to review the status report filed by the Air Force major who had supervised the previous shift. There were three pages, all classified "Secret," and all affirming that nothing unusual had gone on in outer space between the hours of 10:00 A.M. and 6:00 P.M. She made it to the second page before she decided she desperately needed a cup of coffee.

A strong cup helped. She was thinking about the presents lying gift wrapped in the trunk of the car when . . .

"Uh, Commander," drawled an airman sitting at a console in the corner, "maybe you'd like to come over and take a look at this." His voice was steady; it didn't hint of anything. She put down her mug and walked across the room.

The rest of the night, and the next day for that matter, was like nothing Commander Sheila Mondran had ever experienced while working at the United States Space Surveillance Center.

. . .

At just about the time the commander, late and anxious, was descending to Box Nine, nearly a thousand miles south, in the sky above Lake Kickapoo, Texas, the fence was being tripped.

The fence is what the knob turners in Box Nine call the U.S. Naval Space Surveillance System. Stretching three thousand miles across the southern United States from Georgia to California and extending a thousand miles off each coast, this "fence" is actually a man-made energy field that reaches out nearly fifteen thousand miles into space. Transmitters such as the one in the small, south Texas town of Lake Kickapoo send out a continuous fan-shaped wave of radio energy. When an object passes through this beam, the "fence" is tripped.

In the initial half hour after the Lake Kickapoo transmitter detected a bogey, the drill was routine. Receiver stations scattered about the South, from Silver Lake, Mississippi, to Hawkinsville, Georgia, began to "lock in" on the object. Within moments, calculations were made that determined the bogey's speed and size. This information was relayed to the system's headquarters in Dahlgren, Virginia. There the signals were processed by a high-speed IBM 1800 computer that determined the precise position of the object and its anticipated ground track. As soon as this was accomplished, the information was transmitted to the U.S. Space Surveillance Center at Cheyenne Mountain.

There the Box Nine crew took over. Within forty minutes—this was not Flash Traffic; there was no reason to believe the unknown object moving west to east across the southern part of the United States was any cause for concern—sufficient information had been gathered so that a mathematical description of the orbital elements of the object could be fed into the Surveillance Center's computers. This set of working calculations ("We're talking back-of-the-envelope stuff at this stage," a crew member would later explain) included estimates of the period of time it would take the object to complete one revolution around the earth; the inclination to the equator in degrees; the perigee, or just how close it got to the earth; and the highest point of its orbit, the apogee. The preliminary model, now displayed in fluorescent green lines across the maps on their computer screens, allowed the knob turners to confront the next question—what was out there?

Usually, this is the easiest part of the drill. There aren't too many secrets in the space directly above planet Earth. A bogey, after not too much thumbsucking by the analysts, is fitted into one of four categories: it's one of the 7,087 previously inventoried objects in space; it's a satellite whose orbital characteristics have suddenly changed, perhaps as part of a new surveillance mission; it's a decaying object reentering the atmosphere (and then the TIP teams—tracking and impact prediction—get on the job); or it's a newly launched object needing an identification number (though in such cases the Box Nine teams are always notified in advance; nations, continually scrutinized by an array of electronic surveillance sensors, cannot sneak objects into space).

But that December evening the orbital analysts in Box Nine were having problems. It wasn't just that the bogey first targeted by the Lake Kickapoo transmitter didn't fit into any of the usual categories, there was also the problem of its orbital characteristics—they didn't resemble any known operational pattern. There was no military reason for a "bird" to fly like this one.

As Commander Mondran stood looking over the airman's shoulder, what she saw on the computer display didn't seem to make sense. On the screen, the bright green tracking sets were constantly changing. First there would be a lazy, double-helixlike pattern of loops and backtracks, then a set of spiky lines indicating crash dives followed by sudden climbs at astonishing speeds. The object was going through a series of complex maneuvers and rapid changes of inclination at speeds and altitudes that were . . . *impossible.*

But Commander Mondran tried to avoid this conclusion.

"An ELINT?" she suggested to the airman.

He dutifully considered this possibility. In 1984, the Russians had launched Cosmos 1603, a giant electronic intelligence satellite. Tracking that bird had confused Cheyenne Mountain for days; its orbital pattern, too, didn't seem to make sense.

"Could be," said the airman after some pondering.

"Or an ASAT?" she offered. The commander liked this idea. There had been rumors that the Russians had finally developed an effective antisatellite weapon, and perhaps this object, with its unique flight pattern, was a heat-seeking, laser-gyro guided device.

"Could be," the airman repeated evenly.

His indecisiveness infuriated her. "Well," the commander insisted perhaps a little too aggressively, "what do *you* think it is?"

"I don't have to think, ma'am. I'm just an enlisted man."

"*Airman,*" she growled dangerously.

"Well," he finally conceded, "just between you and me, ma'am, it looks like we got someone joyriding up there."

That's when Commander Mondran went back to her desk and picked up the Gold Phone.

"This is the Space Surveillance Center. I have a Flash Alert for CINC-NORAD."

It was a very odd alert. Sheila Mondran had been on duty during two previous Flash situations and those had been steely, tense tours. An awful fear had been sharp and constant—World War III was in the making. But that evening as Box Nine began to fill with command-rank officers and additional orbital analysts, the mood, she observed, was one of puzzlement rather than alarm. It was a night of complications, of mystery. Each newly charted maneuver and change of inclination added to the confusion: military spacecraft—any spacecraft! but no one dared suggest that—do not orbit in such a random flight pattern. What was going on?

And then, just as suddenly as the object had appeared over the southern part of the United States, it disappeared. "Lost him,"; "Can't find it," one knob turner, then another began calling out across the room. Surprise brought their voices to a shout. It was a quick high-pitched chorus.

An Action Message was transmitted from Cheyenne Mountain. The entire Space Detection and Tracking System, a worldwide network of radars, telescopes, cameras, and radio receiving equipment, was ordered to try to lock in on the projected flight pattern of the object.

Simultaneously, the commander in chief of the North American Aerospace Defense Command ordered a Teal Amber search: A computer-linked telescopic system at Malabar, Florida, began to survey the night sky above the southeastern states at a rate exactly counter to the rotation of the earth. It was a technique designed to, in effect, freeze the stars in place and highlight any object that moved across this celestial backdrop. A real-time picture was relayed to the Space

Surveillance Center in Cheyenne Mountain. The resolution was phenomenal; this classified telescope can isolate an object the size of a basketball from 22,300 miles away. Yet it was a long, meticulous search.

The hours passed. Box Nine, as though stunned by the rapidly changing components of the problem, grew quiet. There was a glowing row of computer display screens against one wall, and as the night dragged into morning the attention in the room seemed to fix on these machines. The object would return, and what was real would prove to be rational.

The Space Detection and Tracking System found nothing. Teal Amber found nothing. And by noon—Commander Mondran had been on duty for eighteen hours straight—the only tangible evidence of the existence of the object was the hard-copy printouts the knob turners had made the night before. The alert was still in effect, but by now the search was only for a theory. One popular hypothesis held that the object was a ferret, a solid-fuel, motor-powered homing vehicle that was jettisoned from an orbiting Soviet Cosmos satellite specifically to probe and to confuse the southern air defense system. It was one of Commander Mondran's initial suggestions, however, that garnered the most favor in Crystal Palace: The unidentified flying object had been a Soviet ASAT weapon test, a direct ascent vehicle fired from a fighter plane.

Still, both theories could stretch only so far. There were many unanswered questions. How could the Soviets have designed a spacecraft capable of traveling at such speeds, executing such precise maneuvers, and climbing to such altitudes? Or, perhaps more troubling, did the Russians now possess the Stealth technology that allowed the vehicle to disappear from every Space Detection System radar screen throughout the world?

By the time the alert was officially recalled sometime after three that afternoon, the crew in Box Nine was more concerned about escaping at last from the cave than in pondering unresolved events. Commander Mondran traipsed through the snow-covered parking lot to her car and, mulling over the last twenty-four hours in Box Nine, she made a decision: It's such a crazy, unknowable world; why put off until tomorrow anything she could do today? She would give her boyfriend Jim his Christmas presents as soon as she got home.

And it wasn't until later that night, after the presents had been opened, that Commander Mondran finally got around to telling Jim about the alert. But by then, living once again in the everyday, she recounted the entire episode with a cynical gloss of insider's detachment. She spun a tale, just another one of many, about the fractured goings-on in Crystal Palace. For her, the punch line of the story, the one she and Jim raised a beer to, was an innocuous judgment that might have been any military person's heartfelt gripe—"All that money and all those machines and they still don't know what the hell is happening out there!"

But while in Sheila Mondran's world the events of that night could be dismissed with a lifted glass, elsewhere there were consequences. Cheyenne Mountain Flash Alerts do not go unnoticed. The commander in chief of NORAD had notified the National Military Command Center in the Pentagon. A classified report that an unidentified bogey had penetrated the air space of the continental United States, had executed maneuvers that would seem to be technologically impossible, and then had simply vanished was circulated to the Joint Chiefs and a long list of intelligence agencies. A version of the incident, a tersely factual paragraph, without a drumbeat of conjecture, was even included in the President's Daily Brief. This was the ten-page top-secret document that President Ronald Reagan reviewed each day over his morning coffee. And the paragraph, according to the story relayed to Air Force Intelligence by witnesses, caught his eye. President Reagan looked up from the page and, without a trace of humor, announced that he, too, had once seen a UFO. He had been riding in a plane during his campaign for the governorship of California, and there, right outside his window, was a flying saucer. We should follow up on this, the President said thoughtfully. Still, when no one in the room did any more than nod sagaciously, he let the matter drop.

But as luck and the well-oiled wheels of a constantly grinding bureaucracy would have it, a report on the unidentified bogey also made its way to the E-Ring Pentagon office of the Defense Intelligence Agency's Directorate for Management and Operations. It was there that an obscure Army colonel decided after a day of intense, speculative brooding that it might be elucidating to attempt an experiment. He would ask the participants in Project Aquarius to consider the mysterious sighting at Cheyenne Mountain.

And when he did, a loosely linked chain suddenly began to rattle.

TWO.

As coincidence would have it, the DIA's Directorate for Management and Operations had originally decided to launch its Project Aquarius because of a series of incidents that also had occurred in a windowless room. Only the site for these events had been nowhere near the snowcapped peak of Cheyenne Mountain, but rather thousands of miles away in Washington—in a lead-lined conference hall on the third floor of the ornate Old Executive Office Building directly across from the White House.

The "vault" room belonging to George Keyworth, the President's Science Adviser, was already filled with Office of Science and Technology staff on that day in the fall of 1985—an "eager, opening-night crowd," one who shared the experience recalled—when the two Stanford Research Institute (SRI) scientists announced they were ready to proceed. Seated at a small table at the front of the room was a man the scientists identified only as their "viewer." He was smiling expansively as the older scientist, a big, beefy sort with a sergeant major's bellow to his voice, offered a brief introduction.

What we will demonstrate, the scientist promised, is the existence of a new perceptual channel through which certain individuals are able to perceive and describe remote data not presented to any known sense. This new channel is related, we believe, to the phenomenon known as Remote Viewing. The difference is—and at this point at least one listener thought the scientist sounded somewhat uneasy, as if he, too, had a hard time believing all these extraordinary claims—that while RV requires the viewer to focus his mind's eye on a specific individual, this new channel is employed by focusing on specific latitudinal and longitudinal information.

The scientist then announced, "OK, Commander, we're ready to begin."

A voice from the back of the dark room, loud and authoritative, proceeded to read off a short series of precise geographical coordinates. The scientist, archly polite, now asked his viewer to be so kind as to repeat aloud each of the specific coordinates. The viewer did so without hesitation, which seemed to some a feat in itself.

Then, without another instruction, the viewer became—in an instant—wrapped up in a trance.

Minutes passed. The viewer's head was bowed; it nearly touched the table. The silence in the room was intense. At last he raised his head and spoke softly: "I see a . . . house . . . Big. A mansion, actually . . . Pillars . . ."

When the viewer's words trailed off, the SRI scientist suggested he draw what he had seen. The viewer looked into the distance for a moment, before he picked up his pencil and started to sketch, working quickly and fluidly, like an action painter. He began with geometric forms, rectangles and squares mostly, and bit by bit a pattern emerged.

While he worked, the room was hushed. The only sound was his sharp pencil scratching rapidly across a thick pad of paper. Then, all at once he put his pencil down. He sat with his hands folded in front of him. He reminded one observer of an obedient child awaiting his reward.

"Perhaps you're finished?" suggested the sergeant major scientist.

But there was no answer, just a small indecisive shrug, so finally the scientist took the drawing and began to pass it around the room. He instructed everyone to take a good look, please.

As the crude drawing moved from hand to hand, the other, more youthful-looking scientist motioned to the man in the naval officer's uniform who had been sitting impassively in the rear. It was all done without a word; the scientist, who wore rimless glasses and the longish blond hair of a graduate student, had simply raised his hand as though he were hailing a cab.

The officer rose, and when he walked to the front of the room everyone could see the three gold commander's stripes on his sleeve and the black leather briefcase that was handcuffed to his wrist. The commander—later he would inform Science and Technology staff members that he was assigned to the Office of Naval Intelligence in Suitland, Maryland—quickly unlatched the briefcase from his wrist. Without waiting to be asked, he searched through the case and withdrew a black-and-white photograph. He handed it to the older scientist, who gave the photograph little more than a glance and then, very matter-of-factly, passed it around the room.

The pillared house in the photograph was identical in shape to the sketch that had just been drawn.

Please tell them about the photograph, the scientist, still cool, still detached, requested. The officer was laconic: It had been taken by satellite. He allowed a moment to pass; until, as if gauging the building mood in the room, the commander, with perfect timing, coyly offered a small afterthought—"It's the country dacha of Mikhail Gorbachev."

Only now did the two scientists show some emotion. They beamed at their viewer with the self-satisfied glee of trainers taking measure of their Derby winner. The heavy scientist, bursting with triumph, announced to the room, "And that, gentlemen, is how information is conveyed through the phenomenon we call scannate."

The next phase (or second act, some office wags later joked; for it was all, by design no doubt, quite theatrical) of the demonstration centered on the possible use of psychic powers in antisubmarine warfare. Here again, the commander played the role of the sorcerer's apprentice, though now, clearly part of the show, he was no longer required to assume his bleacher seat. He remained standing at the front of the room, wedged between the two SRI scientists. The viewer, as before, was seated at his small table. He was still smiling amicably at the audience; he might have been an attentive host hoping all his guests were having a good time.

The experiment began when the commander reached into his briefcase and this time withdrew a stack of photographs. Like a man dealing a hand of solitaire, he laid them out one at a time into two short rows directly in front of the viewer. The photographs, perhaps there were six, were face-up, some in color, some black and white. All were photographs of submarines—American and Soviet, both attack class and larger boats armed with land-bound MIRV nuclear warheads.

"I'd like you, please, to study these photographs," the older scientist told the viewer. A further instruction, this time emphasizing each syllable in the word, followed—"Care-ful-ly."

The viewer obeyed. He began at the top row and worked left to right. He stared conscientiously, bending his head low toward each photograph. After a while, he would place his right hand at the stern of the submarine and begin to trace the entire outline of the boat with all five fingers. He moved his hand very slowly, a fraction of an inch at a time. It was as if his fingers were running along the very steel of the boat, appreciating the variety of cold, hard sensations.

It took some time, and all the while, the scientist addressed the audience. He explained that the purpose of this demonstration was to show how scannate might be used to facilitate the detection of submarines. According to the scientist, the viewer (and he spoke as if the man were someplace else; which perhaps was true, so intense was his involvement with the photographs) was at that moment scanning the globe, searching the world's oceans for the whereabouts of these submarines. When he "sees" one, the viewer will describe its location, the scientist explained. He will state the specific geographical coordinates of each site. And how will we know if our viewer is accurate? the scientist asked. The commander, he quickly revealed, just happened to have in his briefcase a chart with the locations of each of these submarines as of twenty-four hours ago.

There was some reaction from the room, and the scientist patiently waited for quiet before pulling the last rabbit from his hat. "One other thing. Just to make this experiment a little more interesting, I've added some controls. Not all the submarines whose photographs are on this table are at sea. One is not even built. Another is in dry dock. Let's see if this throws our viewer off."

Then, again rushing right into it, he turned to the viewer and asked, "Perhaps you would please tell me whether you can scan the submarine pictured in the top row, left-hand corner?"

What was so remarkable, one eyewitness would recall years later, was how the viewer seemed convinced his gift was *un*remarkable. "They might've been asking if he knew the time, for all he cared" was the way this observer put it.

Upon the scientist's command, the viewer would focus on a specific photograph for a beat, then another. When he was done, he would raise his head and zero in on some secret spot in his mind's eye (though as near as anyone present could tell, the place he was concentrating on was somewhere on the rear wall). His eyes would set on this target, until, at last, he would begin a journey. His eyes traveled from some point at the top of the wall to another place near where the wall met the floor. As he did this, his head would gently bob up and down as though he were reading the columns of stock quotations in *The Wall Street Journal*. Perhaps he was "scanning" the globe; it was never really explained.

He moved from one photograph to another without delay. He was very earnest, processing each assignment with an almost robotic efficiency. The sub in the first photograph, a U.S. Navy attack boat, he could not locate anywhere. This didn't seem to bother him or the scientist, so many of the spectators, by now believers, assumed this was one of the jokers the SRI team had put into the deck. And, as the viewer moved down the row of photographs, he appeared to be on a roll: one boat, according to the geographical coordinates he announced, was off the northern coast of Iceland; the next, a Soviet boomer, as subs with nuclear ballistic missiles are called, was at a latitude and longitude that put her in international waters not too far off the coast of South Carolina. Now, this shook some people up. But such uneasiness was only a prelude to what happened when the viewer went on to the fourth photograph, the one at the extreme left of the second row.

At the start, the photograph of this Soviet Delta-class submarine seemed to present little challenge to the viewer. He went into his head-bobbing routine and, as usual, started to call out a set of coordinates. Then he stumbled; his face suddenly became twisted with the surprised look of someone who had just encountered a small, but unexpected trouble in his path.

The scientist caught this, too, and he asked, "What is it?"

The viewer thought for a moment. Was he deciding on an answer? An excuse? The truth? Instead, when he was once more in control, his stride righted, he chose to announce the coordinates. The boat was patrolling a block of ocean between the Maine coast and Nova Scotia.

Yet just as he began scanning for the next photograph, he stopped. His trance broke. All at once he was back among the people in the crowded room, and he was clearly uncomfortable. No, one of the witnesses decided: *he was scared*.

"What is it?" the scientist repeated, also anxious.

The viewer, now a little embarrassed, explained that he had seen something else while he was searching for the last boat. The coordinates were the same, only it was . . . *hovering* above the submarine.

"Was it an airplane?" the scientist suggested.

The viewer groped for an answer. The best he could offer was a shrug.

So the scientist, willing to let the experiment follow its own momentum, asked the viewer to draw, please, what he had seen.

The viewer agreed. Once he reclaimed his trance, he drew rapidly. Again, the sketch had a geometric quality. Circles predominated. Quickly, though, he began to refine it to coincide with his vision. One large circle became narrowed, oblonged. Another circle perched on top of this larger shape was, with a line of his pencil, neatly halved.

It was a drawing of a wingless aircraft. To many in the room, the drawing was quite familiar.

"A rocket?" the scientist prodded after a long look at the sketch.

The viewer heaved his shoulders, a reflexive gesture which could have meant possibly; or, come off it.

The scientist returned to studying the drawing. He considered it quite conscientiously. And yet, finally defeated, he had no choice but to blurt out what was on the mind of nearly everyone in the room. "Well, what else could it be? I mean, you're not going to tell me it's a flying saucer."

"Yes," said the viewer, "that's it exactly."

It was a credit to the scientific objectivity of the SRI team that the classified report on the demonstration in the Old Executive Office Building prepared for its primary client, the Defense Intelligence Agency, was very complete. True, it was only a footnote, but at the bottom of one page, in a note explaining how the percentage of accurate responses (46 percent!) was tabulated, there was a reference to "an untabulated incident" involving the sighting of what the authors of the report referred to as "an unidentified flying object."

And there were many both in and out of Washington who would also doubtlessly insist that it is to the credit of the DIA that the information contained in this footnote, however fantastic, did not diminish its enthusiasm for scanning. In fact, while Dr. Keyworth's aides were left with large, gnawing doubts after a contrived if not prankish demonstration, the DIA was undeterred. Within six months of the experiment in the vault room, the DIA with the cosponsorship of Naval Intelligence (and unknown to the President's Office of Science and Technology) launched a classified operation that employed viewers to scan the globe for Soviet submarines. The operation was called—perhaps pregnantly, perhaps by luck of the computer draw—Project Aquarius.

Over the next fourteen months there were at least seventeen recorded sightings of "hovering unidentified flying objects" by the scanning participants in Project Aquarius. These sightings were well known to the men in the DIA's Directorate for Management and Operations who were supervising Project Aquarius. To some of these intelligence officers, it was ammunition for their cynicism, supporting evidence for their contention that using psychics to scan for submarines was a ludicrous, wasteful experiment. To others, it was a quickly dismissed statistic, a small, unexplained quirk in an exercise that, they were confident, would help to locate Soviet submarines. But to at least one M and O officer these Aquarius sightings would become something more—an inspiration.

Colonel Harold E. Phillips of the DIA was certain he had seen a UFO. According to the story he reportedly had no qualms about sharing with many fellow officers, it had happened nearly forty years ago, on a clear summer's night while walking through an Iowa cornfield with his father. And it had been impossible to forget. The vision of a dome-shaped ship as big as a Greyhound bus and as bright as a Broadway marquee had stayed with him for a lifetime. Ever since that wondrous moment, he told friends without embarrassment, he had been waiting for it to return. Every night, part wish, part reflex, he looked up at the dark sky confident he would finally see it. It was incredible to think so, he knew; but, as he told friends, he was also certain it would be the most natural thing in the world. His long wait haunted him, perplexed him, obsessed him, yet he never doubted that it would return.

Each of the classified Aquarius sightings of "hovering unidentified flying objects" was, the colonel announced to DIA officials, more testimony to a larger and even more dazzling knowledge he had possessed since childhood—something *was* out there. So when the report about the incident at Cheyenne Mountain crossed his desk in the office of Management and Operations, he read it eagerly. Perhaps, he wanted to believe, the moment was at hand—*They* were about to announce themselves. Yet as a day passed and still nothing happened, he grew dejected. Even more troubling, he decided, was the government's, *his* government's, attitude. The President and the Joint Chiefs were allowing a confirmed sighting—and to the colonel's mind there was never any doubt that the bogey monitored by Crystal Palace was anything else but an alien craft—to be filed away as just one more strange, cosmic mystery. Such complacency ate away at Colonel Phillips. He was goaded by it, until he had an idea.

He was able to persuade the scientists running the Project Aquarius teams to conduct a new experiment. The precise latitude and longitude where the fence had first been tripped was provided to three of the most reliable Aquarius viewers. Of course, strict controls were enforced. The viewers were placed in three separate isolation chambers. There was no way any one of them could have known what specific coordinates the others received. Next, they were offered identical instructions: If you can scan for anything unusual at that location within the last forty-eight hours, draw it.

By the end of the day, three sketches were faxed to the colonel's office in the Pentagon. They were all crude, largely geometric pencil drawings. Each was obviously the work of a different artist, but all three were quite similar. They were all rounded, wingless aircraft.

The colonel put the drawings to good use. In that final winter of Ronald Reagan's final term, the DIA was persuaded the time had at last come to convene a top-secret working group to investigate the possibility that extraterrestrials were making contact with this planet—a possibility that for the past four decades the government had publicly insisted was impossible.

THREE.

It was still dark on that February morning in 1987 when the blue government Ford approached the Pentagon's Mall entrance. The driver slowed, expecting the gate to be raised as usual, but this morning was different. The gate remained down and the car had to come to a halt.

A side-armed soldier stepped out of the red brick guardhouse and walked to the car. The military chauffeur, on assignment from the Pentagon motor pool, started to complain, but the soldier ignored him. Instead, he looked into the rear window. There was one passenger in the back seat, a middle-aged man in a suit and tie. A briefcase was resting on his knees.

The soldier stepped to the rear of the car. Taking his time, he compared the Ford's license plate numbers to those listed on the clipboard he was carrying. When he was satisfied, he raised the gate and offered a crisp salute. The driver gunned the Ford's engine as if to make a point, but the soldier held his salute and finally the car sped up the ramp leading to the building's front steps.

When the car was out of sight, the guard made his call. He hurried to the gatehouse and punched in a four-digit number. A phone rang in the Pentagon office of the DIA's Directorate for Management and Operations.

"It's party time," the guard announced, no doubt with some bewilderment; his job was simply to repeat the prearranged code at the appropriate moment—not to understand it.

The DIA official replied with a single question: How many were in the car?

The soldier said there was only one man, and he was the first to arrive.

The DIA official let loose with a loud expletive. And immediately he regretted it. It wasn't the soldier's fault the signals intelligence specialist was early; for an instant, he would recall, he even thought about apologizing. Instead, he clicked off without another word.

The National Security Agency "SIGNIT whiz" was early because he had been able to hitch a ride up from southern Florida in a Navy T-38. The others, as requested, arrived at the Mall gate closer to seven-thirty. There were seventeen men in all. The group included one Army and three Air Force generals, DIA scientists, an Army colonel, three NSA officials, a supervisor from the CIA's Domestic Collection Division, as well as a technical team from the agency's Science and Technology Directorate.

Once past the metal detectors and inside the building, they headed down a long corridor draped with state flags, past the cafeteria, and through the shopping arcade. The bakery smelled like chocolate chip cookies, so it was easy to find. And nearby, as they had been told, was the elevator.

They took the elevator down one flight. These offices beneath the E-Ring were the newest in the Pentagon. A year ago, this basement had been a parking lot reserved for tour buses and taxis, but then military security officers decided the threat from car bombs could not be ignored. Much of the space had been converted into a large surveillance-proof "vault room."

Four soldiers, seated in a row like attentive schoolboys, were stationed by the elevator landing in the basement. On their command, two glass doors were buzzed open. A red-carpeted hallway as narrow as a bowling lane led to a pair of huge metal doors. Two more soldiers were standing at attention on either side of the opened doors. Inside was the Tank, the most secure conference facility in the entire Pentagon complex. The seventeen men filed in.

Nine days earlier each of the men had been contacted by a Colonel Harold E. Phillips, U.S. Army, now assigned to the Defense Intelligence Agency. His invitation had been vague. He had explained that the DIA was convening a new interagency working group to discuss a topic "affecting national security." When pressed he had added, "The area of discussion, in broad strokes, concerns unexplained aerial phenomena." Your presence, he had told them, was not mandatory, but would be appreciated; and, perhaps to emphasize the significance of the meeting, he had added that of course government transportation could be made available "on a priority basis from anywhere in the country." Most of the men thanked the colonel for his invitation, but said they would require a day or so before they could respond definitively. They had to check their calendars.

They did not check their calendars so much as they checked out the mysterious Colonel Phillips. His biography was unimpressive: born 1941, Sioux City, Iowa; graduated University of Southern Illinois, engineering degree and ROTC; entered the Army as a first lieutenant and decided to become a career officer; graduate degree in electronic engineering; various Army signals intelligence postings; and then, in June 1985, transferred to the DIA staff. On paper it was an undistinguished résumé. When they began probing into his work at the DIA, however, they immediately began to reevaluate Colonel Phillips.

He had been assigned initially to the DIA's Directorate for Management and Operations. According to the DIA's organizational chart, Management and Operations was the command directorate for three intelligence groups with similarly misleading, dull organizational titles. But this was just cover. The M and O officers were known to their fellow spooks as this country's premier intelligence analysts. And something more. They also had the reputation for being among the most esoteric; as a CIA official had derisively observed, "You could always count on the M and O boys to be on the far side of far out." Colonel Phillips, it was rumored, had been part of the team that caused quite a stir with the circulation of an infamous (at least in classified circles) monograph, *Psychopharmacological Enhancement of Human Performance—USSR*. This top-secret paper had advocated that in certain carefully chosen situations, American soldiers would, as Soviet tests had already demonstrated, perform more effectively under the influence of selected drugs. Infantry troops, the report contended in one much debated hypothesis, would fight more aggressively after a controlled intake of cocaine; the drug was "a psychopharmacological enhancer of the feeling of invincibility."

There was also another side to Colonel Phillips's work at the DIA. In 1986 he had received a new title: "Associate Coordinator of Space Reconnaissance Activities." He was now the assistant to the DIA's representative to the National Foreign Intelligence Board's Committee on Imagery Requirements and Exploitation (COMIREX). It was one of the most sensitive and powerful jobs in the entire intelligence community.

COMIREX determined the missions for the United States' spy satellites. At a committee meeting, for example, the Navy might request that a KH-11 satellite be positioned over the Polyarny shipyards to photograph a new titanium hull submarine, while the Army might argue that it was more urgent to use the satellite to monitor an ongoing Warsaw Pact tank exercise. COMIREX would decide which request had priority. Also, the committee would determine which of the various intelligence organizations should be on the distribution list for specific satellite photographs.

Clearly, then, Colonel Phillips was a man with an eclectic mind, an intelligence officer cleared, as they say in the trade, "for the world" and, perhaps most importantly, an official too influential to cross. Within forty-eight hours, all seventeen men had phoned his DIA office to announce that they had managed to juggle their schedules. They looked forward to attending the initial meeting of the colonel's working group.

And so early on a bitterly cold February morning, seventeen curious men took their seats around the diamond-shaped, walnut conference table in the Tank. As they greeted one another and helped themselves to strong coffee from a gleaming silver urn, they also exchanged theories about why the mysterious colonel had summoned them. The consensus—a deduction based on the preponderance of Air Force officers, intelligence agency scientists, and the colonel's background in satellite operations—was that this working group would be conducting a Strategic Defense Initiative feasibility study. Has to be something tied to Star Wars, everyone seemed to agree.

They were all wrong.

They realized this just moments after a short man with a graying crew cut and a faint mustache strode to the lectern at the front of the room. "Gentlemen," he began, "my name is Colonel Harold E. Phillips and the subject of this morning's briefing is the events that have led to the formation of this committee. By the way, henceforth we will be known as the Unidentified Flying Object Working Group."

Then, ignoring the bewildered murmurs, Colonel Phillips reached into the lectern and removed a remote-control device. A button was pressed and automatically the lights in the room dimmed, while a white movie screen was lowered from the ceiling.

For the next forty-five minutes, as a series of slides provided by the U.S. Space Surveillance Center clicked into focus, the colonel spoke without interruption. He gave the details of the mysterious Cheyenne Mountain alert and shared the history of Project Aquarius. When he reached the point in his narrative where he revealed the most recent experiment the Aquarius scientists had run, three sketches were flashed in rapid succession on the screen. They were the pencil drawings of the domed, wingless craft the viewers had scanned at the precise location where the fence had been tripped.

It was at that moment that an Air Force general broke in. "That's like nothing we've got," he said.

The colonel nodded his head in agreement and added, "Or the Russians. At least that we know about."

Suddenly an Army officer spoke up. He seemed annoyed, and his tone was, it would be recalled, loud and aggressive. He demanded to know if Colonel Phillips thought the object that had been monitored by the Space Command was an alien spacecraft.

The colonel waited before answering. A white light shone intensely on the black screen. When at last he spoke, each word was articulated carefully, as though after much consideration.

"Sir," he said, "I think we all have to agree that's a distinct possibility."

FOUR.

Once the unspeakable had finally been spoken, a new tenseness filled the Tank. There was quiet, but it was a controlled silence. Everyone was holding back. The men seated around the conference table, alert, some even thrilled (or so they would later admit), focused all their attention on the man at the lectern.

Colonel Phillips seemed impervious to their gaze. Now that he had made his speech, shared his secrets, he appeared almost indifferent. Either you believed him, or you did not. Either you accepted the irrefutable logic that tied together the events he had outlined, or you missed the point. The choice was yours. When he finally spoke again, his tone was muted. There was an air of self-effacement—and, also noted, a quiet, convincing sincerity—to his entire demeanor. He began with a small question. He asked if anyone present could offer alternative explanations for the sighting at Cheyenne Mountain and the Project Aquarius drawings.

If anyone could, he chose not to share it. Still, this silence seemed neither to encourage nor discourage the colonel. He acknowledged it with only the briefests of nods and then went on. The DIA, at his urging, he explained, had decided the time had come for responsible officials of the United States government to investigate, without preconceived prejudices or institutional constraints, the whole range of questions raised by the possible existence of life on other planets.

And, the colonel continued evenly, this Working Group had been assembled with the hope that it would agree to supervise and direct this investigation.

At this point, one of the men sitting at the table interrupted. He asked a question about "all those kooks" who were "always seeing flying saucers." He wanted to know if the colonel thought such reports were "credible." Did the colonel, he asked, think these civilian sightings should be examined by the Working Group?

The colonel, to the amazement of at least one person in the room, answered, "Where there's smoke, there's often fire." And, deftly, he reached for another cliché to embellish this notion. "After all," he declared, "even paranoids have enemies."

From around the table there were, it was observed, nods of agreement; conceivably this was just what the colonel had anticipated—after all, he was addressing a roomful of men who had served lifetimes exploring the boundaries of their suspicions.

Abruptly, however, this communal mood was broken with one word.

"Rubbish," bellowed a man who was later identified by a participant as a longtime veteran of the CIA's Directorate of Science and Technology. He rose from his seat as he continued. If this Working Group was ever going to accomplish anything, he said, it would have to approach the phenomenon of unidentified flying objects "with science, not science fiction." It wasn't good enough to rely on reports from the wild-eyed and the fanatic. We must rely on science.

"Science, science, science," the CIA man repeated, near to singing. Science will have the answers. Science must analyze the clues. The Working Group can on demand have access to "the best and the brightest" minds in all fields. We must use these resources, we must question them, we must confront them, he insisted with a fervor suitable to a halftime locker-room oration. And then, looking slightly embarrassed, he sat down.

Colonel Phillips again took center stage, while a roomful of men familiar with command waited to weigh his response.

"Thank you," he said in an earnest and pleasant voice to the CIA man. "I agree." And, to the further surprise of some of the men sitting at the table, he announced that the goal of the Working Group, as he saw it, was not simply to investigate the UFO question, but also "the question of whether or not the human race is alone in the universe." "Our task," he offered, "is a scientific one—to search for proof of extraterrestrial intelligence."

Yet just at the moment when some of the men in the room were silently appreciating this small triumph of diplomacy, there was another challenge. A voice called out softly, "We won't find anything."

The speaker was wearing a dark suit and a correspondingly somber tie and he explained that before he had joined the National Security Agency he had been trained as a physicist. His tone was passionless, almost flat; it seemed to signal a sudden drop in the air pressure after all the mounting excitement. He went on: "I don't believe we'll discover one blessed thing. We are alone in the universe."

Colonel Phillips was about to interrupt when the man restrained him with one word—"Please." The colonel immediately apologized, and the NSA official picked up where he had left off.

But, he said, and it was now clear that the spring of his conviction was winding tighter, "I support the Working Group and its mission one hundred percent." His rationale, as he proceeded to share it, was passionate. If it can be demonstrated once and for all that the human race is the only inhabitant of this infinite universe, then how much more precious that makes each and every one of us, how much more profound that makes our responsibilities. He concluded with nothing less than a plea. It is the duty of those on planet Earth to assume their destiny as the sole explorers of the universe.

After that there was no more debate. If there had been any doubts, they were now suspended. The unique responsibility of the UFO Working Group was acknowledged by all in attendance. The mood in the room swung toward the self-congratulatory: We, of all earthlings, are the chosen.

It was an exciting time; and it was just the morning of day one.

Lunch was served in the Tank. A soldier wheeled in a cart that held two trays loaded with sandwiches, a fresh urn of strong Pentagon coffee, and a stack of china plates embossed with the seal of the United States of America. Colonel Phillips genially told his guests to help themselves. He also, a bit more pointedly, offered one further instruction. They were not to discuss the morning's briefing during the break. The reason, he explained, was to ensure that the afternoon session would be "spontaneous." After lunch, he promised, there would be "ample time for candid discussion and debate."

The meal, under those restrictions, was a quick affair; and as soon as the dishes were removed, Colonel Phillips called the meeting to order. He had abandoned his lectern and was now sitting at the conference table. The change of situation seemed to transform him. He was no longer hesitant, no longer dispassionate. Proximity to the other men had energized and emboldened him; or, like many men in intelligence work, perhaps he simply adopted moods to complement whatever the circumstances required. Without even a brief introduction, he started in reading a "statement of purpose." His new pitch was loud and booming; clearly, it suited him to leave little doubt he was in charge.

The statement was presented in concise, enumerated paragraphs and, according to the memories and notes of two government officials present, it declared: One, the Unidentified Flying Object Working Group was to review and evaluate the information obtained by all previous official inquiries, both public and clandestine, into the UFO phenomenon; two, the Working Group was to review and evaluate all attempts by the scientific community to search for extraterrestrial life; three, the Working Group was to review and evaluate selected sightings and other evidence of UFOs as alleged by private individuals; four, the Working Group was at all times to keep its existence secret from both government agencies and the public; and five, its institutional life was to be open-ended, as long or as short as was necessary to determine positively the existence of extraterrestrial life.

The "statement of purpose"—daring, exhaustive, and determinedly clandestine—was adopted that first afternoon by a unanimous voice vote.

The UFO Working Group's mission to determine whether the human race was alone in the universe could now begin.

Six months after that secret meeting, I began to ask my first, cautious questions about the UFO Working Group. From the outset, there were problems. No one was willing to admit to being a member of the panel. I made long lists of possible committee members—all the usual suspects and then some—and spent hours trying to arrange interviews. Many officials refused to talk at all. Those who gave in adamantly denied any knowledge of the UFO Working Group.

And I knew they were lying.

The government's intention, I was becoming convinced, was not simply one of concealment. Rather, its purpose was to jostle any investigator who was asking questions about its ongoing covert search for extraterrestrial life, to keep him discouraged, misled, and confused. Goethe's famous wisdom, I came to realize, was a half-truth. Doubt was not the only thing that grew with knowledge. So did suspicion.

I was not a believer. I was only a reporter chasing a story. But I had to wonder— Why? Why all the lies? What was the government trying to hide?

Finally, I was put in touch ("put in touch": three circumspect words will have to summarize the more complicated story it would be a betrayal to tell) with an intelligence official who had attended the founding Working Group session. We developed, as it is called with deliberate coyness in my chosen trade, a working relationship. Soon, another source shared a bit of vital information, and then decided, quixotically, to withdraw. But I kept at it; the lies and obfuscations gnawed at me. And, in time, my doggedness brought reward. One more Working Group member, a witty and perceptive veteran of the intelligence community, was willing to share his firsthand observations. I now had two sources on the inside of the UFO Working Group.

But their cooperation had its limits—and its rules. They would not allow their names to be used. They would not, in fact, disclose any of the names of the individuals chosen to serve in the Working Group; I would have to make do with institutional affiliations. And, one more nondebatable topic, they would not provide me with any minutes or reports issued by the Working Group; that would be a violation of not only their principles, but perhaps also of federal espionage statutes.

Still, those two men independently wove quite a story. They seemed, now that they had made the decision to cooperate, to enjoy talking about the UFO Working Group and its activities. I was becoming confident that, if my luck and ingenuity held, I might be able to uncover the entire history of the Working Group. But when I attempted to track down Colonel Harold Phillips, I once again was thwarted—and this, too, left me wondering.

The first time I called the DIA office in the Pentagon, an efficient voice informed me, "Colonel Phillips is not available. Would you like to leave a message?"

I did. It was the first of many. All went unreturned. It became a bit of a game. I took to calling the colonel's office at odd hours. Nine in the evening. Nine in the morning. Whenever I had a spare moment.

At last, early one morning, the colonel took my call.

"You've been trying to get in touch with me, I gather," an unexpectedly high-pitched voice declared. "Well, what can I do for you?"

I was caught unprepared. After some hesitation, I managed to mumble that I wanted to ask about the UFO Working Group.

"I see," he said.

While I waited, I was gripping the pencil in my hand with all my might. I was prepared to write down his every word.

"Never heard of it," the colonel finally said. "Now if you'll excuse me, there's some work I must return to."

The next time I called the same Pentagon number, I was told by the efficient voice, "I'm sorry. We have no record of a Colonel Harold Phillips."

But by then it didn't matter: I was already on to him, and the UFO Working Group.

Part II
THE BURROWERS

FIVE.

It is one of the operational wisdoms of intelligence work that a shrewd burrower—the deskman who can resourcefully dig his way through a pile of files and documents to uncover the single relevant fact, the one missing clue that long ago had been carelessly ignored or foolishly discarded—is worth his weight in field men. And so it was not entirely unexpected when I learned that one of the first activities of the UFO Working Group was to review the government's mountains of abandoned UFO files. The eager committee, however, did not recruit a team of burrowers. Instead, it saved time by summoning a man who just three years earlier had ordered his own covert team—now *this* I had not anticipated—to make its careful way through these same dusty, official archives.

And, I discovered, this call to address the UFO Working Group had also taken Major General James C. Pfautz, U.S. Air Force, Ret., by surprise. Like a winning lottery ticket, it had been hoped for, dreamed of, yet never really expected. General Pfautz couldn't believe his luck.

Ever since his retirement as chief of Air Force Intelligence, after nearly a lifetime of being on the inside, Jim Pfautz had been trying to get used to being a civilian. He had forced himself to stay away from the Pentagon, to putter around his house in Alexandria, to get his news straight from *The Washington Post* and *The New York Times* and never succumb to the gnawing impulse to pick up the phone to find out what was really happening. He had tried to convince himself his new life was busy and full, a well-earned chance to travel with his wife, to lean back and enjoy things. Yet, he realized, all his efforts went toward a losing battle. Each day was lived with a struggle to reach a sort of philosophy: No military man is indispensable.

But now they were calling him back! The summons from the UFO Working Group on that March day in 1987 was, he convinced himself, his opportunity—a passport back into the world of secrets. Who knew where it could lead? It amused him that such a small, inconsequential part of his intelligence career was bringing him back into the loop, but he was too cautious to share the irony. What was that saying, he told himself, about a gift horse? Besides, before his departure, his team of burrowers had made progress through the files. He did have a story to tell the Working Group. And, he also reminded himself, his exit from the corridors of power had been too precipitous. He deserved to be back on the inside.

Four years earlier, in 1983, Jim Pfautz, I learned, had been riding high. Just fifty-two, with West Point, 188 Vietnam combat missions, and a successful tour as chief intelligence officer in the Pacific command behind him, he seemed the "hot runner," the man to watch. With his second star new enough to gleam, he took command of the more than 33,000 Air Force Intelligence operatives that spring determined to shake the eggheads up. Those *intel weenies*—ex-fighter jocks like Pfautz never had much truck with guys who only had flown a desk—were not going to sidetrack his career. No more dismal, ignorant shrugs; no more dead-fish voices apologizing, "Sir, I don't know the answer, but I'll find out"; no more pencil-sucking colonels trying to weasel out by suggesting, "Maybe you should run that by the NSA" or—unforgivable—"I'll check to see if the Navy guys at Suitland have anything on that." Let them call him Steve Canyon behind his back, but Jim Pfautz was not going to back down: Air Force Intelligence would become the best military intelligence organization in the world. What AFIN didn't know, it would find out. On the double.

It was an exciting time. Being on the inside, having the government's deepest secrets delivered to him at his command was the trigger for adventure. Each day there was something new. The classified folders arrived on his desk brightly color-coded like the pieces in a toddler's jigsaw puzzle: red was KEYHOLE, or overhead surveillance information; blue was HUMINT, the pickings from the agents in the field; and white was ALL-SOURCE, insights processed from a variety of resources. Each day a tangled, turbulent planet was scaled down to size. The world was suddenly just another chessboard, and he was a ranked player.

The first lesson of intelligence work came easily: Knowledge is power. If there was anything Jim Pfautz wanted to know, it was his for the asking. He was not tentative. As head of AFIN, he apparently felt he had a *right* to know.

Jesse Jackson was a case in point. After a routine briefing about Jackson's starring role in the successful negotiations for the release from a Damascus jail of Navy Lieutenant Robert O. Goodman, Jr., a navigator who had been shot down over Syria in late 1983, Pfautz decided to make further inquiries. Perhaps Jackson's diplomatic grandstanding just didn't sit well with the general. It wasn't seemly, especially if the man hogging the limelight was a minister. And that's when it occurred to him—maybe Jesse Jackson hadn't been ordained after all. Maybe there was some long-hidden scrap of information about presidential candidate Jackson's ministerial past, a secret tucked away in a dusty corner of his congregation in Chicago, that Air Force Intelligence could discover—and use. So an AFIN lieutenant colonel was dispatched to Chicago. After days of playing field man, the spy came in from the Windy City with his notebook still empty; he had discovered nothing out of the ordinary. But that was almost irrelevant. Two equally vital objectives of the exercise were boldly demonstrated: the ubiquity of Jim Pfautz's curiosity and the blind obedience of his team.

Yet while his scope was capricious, the general was aware his was an Air Force service. His natural domain was the wild blue yonder—and its mysteries. In 1983, six months after taking command of AFIN, he decided to investigate the heavens.

To Pfautz's way of thinking, the decision to establish this secret UFO task force was not that extraordinary. The primary job of the Air Force of the United States of America was to protect and defend the airspace of this country. No intrusions could be tolerated. It was his job, he reasoned, to investigate whether unidentified objects of any sort, of any origin, were penetrating this airspace. He was simply fulfilling the responsibilities of his office.

The cost of this UFO task force, however, was a problem. Not that there was a lack of funds. Air Force Intelligence (which included the National Reconnaissance Office) was the biggest and richest spy organization in the entire intelligence community. Its 1983 budget was at least three billion dollars. But there was a catch: Congress had to be informed of any expenditure over three million. And there was no way General Pfautz was prepared to go hat in hand to the Hill asking for five million dollars for UFO research.

It was peanuts, a small sum of money to be spent on a vital national security issue, but the general knew he couldn't count on Congress or the public having the wisdom to appreciate what was at stake. They would brand him the resident Pentagon loony, General E.T., and that was no way to go about getting a third star. Or worse, the story would leak to *The Washington Post* and every nut who had been to Venus and back would be pounding on his door.

The general was convinced he had no choice but to take the operation black. He decided to run it as an "off the books" show. The projected five-million-dollar budget for UFO research would be split three ways—and a skeptical, publicity-seeking Congress would never have to be informed. Only now there was another problem. Which of the other players in the spook community would be willing to ante up $1.67 million, more or less, for a chance to tag along and maybe—it was a *big* maybe—share the intelligence product of the century?

The National Security Agency was one logical candidate. The NSA, officially in the business of eavesdropping, was an always curious, electronically sophisticated organization; the prospect of focusing its surveillance sensors toward space might be an appealing challenge. And the agency certainly had deep pockets. While the actual dimensions of the NSA budget are scattered and secret, its annual allotment is at least two billion dollars. The only trouble was, General Pfautz was in the midst of a long, running feud with the NSA director. Years earlier, Pfautz had criticized the security procedures of an operation under General William Odum's command in West Germany, and the stony, intense NSA director had a long memory. He had no friendship for Jim Pfautz. The NSA, as a pragmatic consequence, was never approached. Likewise the CIA. It was hard to imagine the Company providing more than a million dollars and then sitting back and letting the Air Force run the show. It was partly by the process of elimination, then, that the Air Force decided to approach Army Intelligence and the Defense Intelligence Agency.

In the early 1980s, Army Intelligence was an inventive, why-not-give-it-a-try sort of organization. It was spending millions on parapsychological experiments. It had contracted with the Monroe Institute in Faber, Virginia, for studies to relieve stress through "advanced states of consciousness." It had spent research dollars on "hemisphere synchronization," a process that uses patterns of stereo sound waves to intensify consciousness by "uniting" both hemispheres of the brain. Marksmen were being taught to concentrate through paranormal methods (a project so intense that several officers later claimed they suffered recurring mental problems as a result). And the commander of INSCOM, as the Army's Intelligence and Security Command was known, was Major General Albert Stubblebine, a man nicknamed "Spoon Bender" because of his rumored belief in psychic powers. When the Air Force gave this brother intelligence group the chance to join the hunt for UFOs, the Army—and its money—came on board without hesitation.

The Defense Intelligence Agency was approached for somewhat different reasons. The DIA had been established by Secretary of Defense Robert McNamara in 1964 to serve as the central coordinator for all military intelligence. As McNamara had conceived it, the DIA would present the Joint Chiefs with a comprehensive intelligence product shaped from the shared resources of the Army, Navy, and Air Force. Interservice rivalries would vanish and be replaced by a spirit of cooperation. It was a wonderful concept. It just never came close to being realized.

By the time General Pfautz was commanding Air Force Intelligence, the DIA was in shambles. It had grown into a large, unwieldy organization rutted with independent intelligence fiefdoms and esoteric study groups. And, compounding the problems, its head, Lieutenant General James B. Williams of the Army, didn't seem to possess either the energy or the desire to sort things out.

In some ways, however, all this suited General Pfautz. He wanted to head the DIA. It was Pfautz's large ambition that his accomplishments as head of AFIN would one day propel him into Williams's job; then he would grab the DIA by the scruff of the neck and show the Joint Chiefs how a professional runs an intelligence agency. In the meantime, he was willing to let the DIA join his UFO task force. As for the DIA, it was game for just about any interservice intelligence program. It, too, signed on eagerly.

But then, after the players had been recruited, after the checks had been deposited, after a team of burrowers had been assigned to sort through a library of classified UFO documents in an attempt to discover just what the government, for forty years publicly dismissive and exasperated, *really* knew about UFOs, Major General James C. Pfautz made the mistake that was to end his career.

Following the September 1, 1983, downing of Korean Air Lines civilian jet 007 by a Russian interceptor, General Pfautz and his AFIN backbeaters went to work to sort out the bottom-line reasons for the shooting. Was it an accident, or had the Soviets deliberately and knowingly targeted a jet with 269 civilians on board? The general's finding, set out in his customarily blunt, no-nonsense terms, came down to this: The Soviets had believed they were shooting down an American reconnaissance plane—one had earlier been in the vicinity—that had provocatively violated their airspace.

It was the wrong theory.

It wasn't so much that it was faulty; Pfautz's hypothesis had its share of supporters. But what was so unforgivable in official Washington was the general's vociferous allegiance, within the walls of the Pentagon, to an analysis that was directly contradictory to government policy; both the President and the secretary of state, after all, had been quick to announce to the nation that the shooting was further barbarous proof of the existence of an evil empire. And as the general continued to challenge the distortions echoed by Pentagon loyalists, he was passed over for his third star. He was never promoted to director of the DIA. In 1985, reading the signs, he retired. His UFO task force was only one of many ambitions that were retired with him.

And then, two years later, I learned this new UFO Working Group had called him back. They needed his memory. They wanted to know what pieces in a vast, confusing, and often deliberately misleading puzzle—two generations' worth of official secrets about UFOs—his burrowers had already discovered.

At precisely 8:00 A.M. on a March morning in 1987 Colonel Phillips stood at his lectern in the Tank and announced that former Major General James C. Pfautz would address the Working Group. The general's topic: A History of the Military's Involvement with UFOs.

SIX.

Regrettably, no transcript of General Pfautz's remarks exists, which was perhaps what the general, always protective of his secrets, had intended. My only alternative, then, was to roll up my own sleeves, to become, in effect, one of his burrowers digging through the forty-year treasure trove of official UFO documents. And so, armed with the Freedom of Information Act, guided by the wisdom of cooperative sources in the intelligence community, I went off to ferret through the evidence hidden away in government reading rooms. I would blow away the cobwebs and dust off yellowing files as I searched for an answer to the same tantalizing question that had challenged the general's team: What did the military really believe about UFOs?

But where should I enter this maze? There were intimidating walls of official papers, each a turn in a new direction, each a potential clue. It was a CIA official who suggested that one starting point might reasonably be the seminal CIA monograph, *The Investigation of UFOs*. It was an attractive point of embarkation because not only was it a comprehensive chronological survey, but it also was a genial read. Though graded "Secret" at the time of its publication in 1961, the paper opened with a narrative charm suitable to a fairy tale:

". . . In 593 B.C. Ezekiel recorded a whirlwind to the north which appeared as a fiery sphere. In 1254 at Saint Albans Abbey, when the moon was eight days old, there appeared in the sky a ship elegantly shaped, well equipped and of marvelous color. . . ."

This stylish tone, however, turned abruptly feisty and judgmental as the story reached "the modern era of UFO's." On June 24, 1947, the monograph continued, Kenneth Arnold, a fire equipment salesman piloting his own plane near Mount Rainier in Washington State, observed nine luminescent objects flying at extraordinary speeds across the sky. Arnold, frightened and bewildered, described each of these odd vehicles as resembling "a saucer skipping across the water." The wire services latched onto the story; and, as so often is the case, an anonymous autocrat on the copy desk wielded a fateful red pencil. When the papers hit the stands that evening, the unknown craft had become "flying saucers." "A new epoch," the CIA proclaimed, "had dawned."

Or had it? For, despite the CIA's self-assured pronouncement, perhaps that truly epochal event in the long history of the government's fascination with UFOs did not occur, I reasoned, until three weeks later. On July 12, 1947, the commanding general of Wright Field, Dayton, Ohio, received a written account from Kenneth Arnold asking for "an explanation of these aircraft . . ." The letter had the urgent feel of a telegram since Arnold, in haste or for emphasis, took to omitting articles and conjunctions from his sentences. Still, it was undeniably earnest: "It is to us very serious concern as we are as interested in the welfare of our country as you are." And it was resoundingly effective. The commanding general of Wright Field started a file.

In the beginning was the file. Once an official Air Force file was created, reports of unidentified flying objects began to pour in from around the country. Where there had once been a bureaucratic vacuum, there was now a world that was quickly being populated with sworn statements, photographs, memos, news clippings. Week by week, the stacks of unsolicited information in the Air Force's possession grew. There was no specific program to handle this flow of information, yet it all kept on coming. Its heft alone was formidable. It could not be ignored. It demanded attention.

"To end this confusion," the CIA monograph observed, "the Chief of Staff directed on 30 December 1947 that a project be established to collect, collate, evaluate and distribute within the government all information concerning sightings which could be construed as of concern to the national security. Responsibility for the project, assigned the code name 'Sign,' was given to the Air Technical Intelligence Center."

Project Sign looked at 243 sightings and in February 1949 submitted its findings:

"No definite and conclusive evidence is yet available that would prove or disprove the existence of these unidentified objects as real aircraft of unknown and conventional configuration. It is unlikely that positive proof of their existence will be obtained without examination of the remains of crashed objects. Proof of non-existence is equally impossible to obtain. . . ."

So much for the CIA's ambitious hope "to end this confusion." And from the swamp of official ambivalence crawled out, fully grown, another government task force. Project Sign, with an almost biblical immediacy, begat Project Grudge.

The Grudge Report, despite its combative name, was a low-key evaluation of 244 sightings. Its conclusion was simply to shrug off all the nationwide concern as just so much silliness: ". . . reports of UFOs were the result of misinterpretations of conventional objects, a mild form of mass hysteria or war nerves, and individuals who fabricate such reports to perpetrate a hoax or to seek publicity."

The authority of Project Grudge, however, was short-lived. The publication of its slim report had only a single unequivocal result: the formation of another official commission dedicated, or so the press announcement promised, to the preparation of a *truly* once-and-for-all final and conclusive analysis of the UFO phenomenon. In March 1952, Project Grudge begat Project Blue Book.

I became a prisoner of the reading room. The eyewitness reports collected by these first government task forces were a tour across a concerned and troubled America. Making my way through the files, I encountered four Texas Technical College professors who observed two strange formations like "strings of beads in crescent shape" zoom across the Lubbock, Texas, night sky. Then there was George F. Gorman, a construction work manager in Fargo, North Dakota, who, according to the Sign file, "did not make the impression of being a dreamer . . . reads little . . . spends 90% of his free time hunting and fishing" and was certain he had seen "a small round ball of clear white light." And Sergeant Quinton A. Blackwell, who while on duty as chief operator in the control tower at Godman Field, Fort Knox, Kentucky, observed "a metallic object of tremendous size." Intrigued, I read on and on.

Each sighting affidavit was its own unique drama. Some were sincere, poignant in their intense desire to be believed. Others, though remarkably vivid, were wild, desperate hoaxes, the stuff for headlines in those newspapers that are clustered around supermarket checkout counters. But, regardless of how credible the evidence, the government's public attitude remained stolid. In the decade following World War II, when citizens throughout the country were sharing their fears with the Sign and Grudge investigators, the government maintained an official position of adamant skepticism.

Yet, it was all an act.

For I discovered a fork in this paper trail. A world of once classified UFO files, like a parallel universe, existed—and these documents told another, more anxious story.

The pattern of behind-the-scenes maneuverings had been set in motion from the start. While Project Sign proceeded out in the open, scrutinized by both the press and concerned citizens, the Air Force conducted its own secret investigation. In 1948, the Air Technical Intelligence Center ordered a classified "Estimate of the Situation."

According to Captain Edward Ruppelt, who would later head the Air Force's Project Blue Book, Estimate investigators had reached a definitive conclusion: the unidentified flying objects were of extraterrestrial origins. But General Hoyt Vandenberg, then chief of staff, immediately rejected the draft document for "lack of verifiable proof." On the general's instructions, Estimate was burned before it could be distributed.

Instead, Air Force Intelligence Report No. 100–203–79, "Analysis of Flying Object Incidents in the U.S.," was issued. This twenty-six-page document, though less dogmatic in its conclusions than Estimate, was a contradiction of the Air Force's public stand. It was classified "Top Secret" and it sounded a warning:

"It must be accepted that some type of flying objects have been observed, although their identification and origin are not discernible. In the interests of national defense it would be unwise to overlook the possibility that some of these objects may be of foreign origin."

More ominously, this classified report also raised the possibility that these craft were a danger to the security of the United States:

". . . if it is firmly indicated that there is no domestic explanation, the objects are a threat and warrant more active efforts of identification and interception."

This fear of the intentions of these unidentified objects, this possibility of "a threat" to the security of the United States, was to be repeated over and over in classified documents. It became a chorus of official anxiety and, in time, a justification for secrecy. A confidential Air Force Office of Special Investigations report looking into the sightings of "fireballs" near the Los Alamos, New Mexico, nuclear research facilities was typical: ". . . the continued occurrence of unexplained phenomena of this nature in the vicinity of sensitive installations is cause for concern."

Turning the pages of these once secret documents, I found it was almost possible to hear parade-ground minds grinding away: Even if there was nothing conclusive to know, there still was no need for the man in the street—for his own good, of course—to learn the entire, nebulous story. Civilian panic was to be avoided at all costs. From the start, it was to be an insiders' game.

This was not just an attitude; it quickly became the rule. Air Force Regulation 200-2 issued on August 12, 1954, and signed by General Nathan Twining, chief of staff, concluded with an emphatic instruction under the heading "Release of Facts":

"In response to local enquiries, it is permissible to inform news media representatives . . . when the object is positively identified as a familiar object . . . For those objects which are not explainable, only the fact that ATIC [Air Technical Intelligence Center] will analyze the data is worthy of release, due to the many unknowns involved."

There were no lies, just necessary deceptions. And no harm—the Air Force was on top of things. UFOs were *its* job, and it remained vigilant. On December 24, 1959, base commanders around the United States received an official reminder:

"Unidentified flying objects—sometimes treated lightly by the press and referred to as 'flying saucers'—must be rapidly and accurately identified as serious USAF business in the ZI [Interior Zone] . . . Technical and defense considerations will continue to exist. . . ."

The UFO mystery had become by the end of the Eisenhower era "serious USAF business." The public would be fed Project Grudge, with its facetious, bemused explanations. Candor, it seemed, was not in the national interest. I realized that as, accompanied by a former

Air Force intelligence officer, I sat in a dark screening room and watched the consequences of overstepping those bounds.

In January 1958, the CBS television network had devoted a live episode of the "Armstrong Circle Theater" show to a discussion of UFOs. Throughout most of the broadcast Marine Major Donald Keyhole, a believer, was battling two Air Force spokesmen. Not that there was much the major could do; the parameters of the debate had been constricted before the show got on the air. The script the major was being allowed to read from the TelePrompTer was a heavily edited version of what he really wanted to say. He could only offer vague suggestions to the American people, nothing like the headlines he was convinced of, and each sentence gnawed at him. Suddenly, he threw it all away. The whole nation was listening and he was sounding the alarm: "And now I'm going to reveal something that has never been disclosed before . . . for the last six months we've been working with a Congressional committee investigating official secrecy about UFOs. . . ."

And then the audio went dead. The major kept on speaking, but it was impossible to hear a word he was saying.

The CBS director of editing offered an explanation after the broadcast: "This program had been carefully screened for security reasons. Therefore, it was the responsibility of this network to ensure performance that was in accordance with predetermined security standards."

And so, the policy had been set and accepted by all levels of the establishment. Need to know, compartmentalizations—those were the marching orders of the day. The circle of official secretiveness grew even tighter in the 1960s. Defense Department Assistant Secretary Arthur Sylvester, in fact, thought nothing of flaunting it. Withholding news from the public concerning UFOs, he announced to the press on October 29, 1962, was justified if the ends necessitated it. He cited as his rationale Air Force Regulation 11–7: In certain situations information requested by Congress may not be furnished "even in confidence."

It was at this point, after months of weary-eyed reading, that I found myself recalling with some amusement a line in the secret CIA monograph that had been my starting point. *The Investigation of UFOs* huffily complained that in the twenty years following Kenneth Arnold's first sightings "it became fashionable to accuse the Air Force of censorship or of withholding UFO information from the public." And now I could understand why. For it was all preserved in the official files; they were a monument to the birth of a long and abrasive era of national skepticism. One day I would read military task-force reports proclaiming there was no need for concern, no unanswered questions about UFOs. On the next, the Freedom of Information Act would unlock padlocked cabinets stuffed with secret trails, swarming with leads and false leads, classified documents pregnant with lingering problems. In those memos written for a restricted readership, there was always something—a fact, an ambiguity, an opinion—that was deliberately held back from the public. A succession of generals, CIA administrators, FBI officials, and presidential advisers had believed that their investigations should move forward behind the scenes. To tell all, their attitudes and actions suggested, would be a sin, a betrayal of institutional convenants. A fundamental law of spook behavior, as inviolable as an axiom of physics itself, guided the bureaucrats: For every story shared, there was one buried; for every truth announced, there was one suppressed. And so, perhaps inevitably given the cast of official characters involved, given their fears and concerns, from the first a cover-up had begun.

But what were these officials really hiding? The early classified files revealed that an anxious government had deliberately misled the American people. Yet, one large question still remained. Did the government in the decades immediately following World War II have proof of the existence of extraterrestrial life? Or, ultimately, was it as ignorant as any man in the street?

With mounting interest, I read on.

The Air Force's Project Blue Book was the most ambitious of the official UFO investigations. The inquiry had lasted for seventeen years; its files were voluminous. Blue Book officers had been assigned to every Air Force base in the nation. For help on suspected aircraft sightings, the Blue Book investigators had gone to Federal Aviation Administration offices, local airports, the Strategic Air Command (SAC), and the Air Defense Command. In suspected balloon sightings, assistance had been provided by the U.S. Weather Bureau, the Holloman AFB Balloon Control Center in New Mexico, the National Center for Atmospheric Research, General Mills, Raven Industries,

Sea Space Systems, and universities. For suspected satellite sightings, the investigators had consulted the printed Echo satellite schedules, National Aeronautics and Space Administration (NASA) satellite reports, the Smithsonian North and South Equatorial Crossings, and the Space Detection and Tracking System. For suspected missile observations, the Blue Book analysts had conferred with authorities at Cape Kennedy, Florida; Vandenberg AFB, Lompoc, California; Point Mugu Naval Air Station, California; Wallops Island, Virginia; Eglin AFB, Florida; Holloman AFB; and Green River. Photographs had been sent to the CIA's National Photographic Interpretation Center and to the Eastman Kodak Company. Physical specimens had been shipped to the Air Force Materials Laboratory, Battele Memorial Hospital, the Food and Drug Administration, Libby-Owens, Corning Glass, the Institute of Paper Chemistry, and the Geology Department of Northwestern University.

More than 13,000 cases had been examined, and in 1969 the study found:

"(1) No UFO reported, investigated, and evaluated by the Air Force has ever given any indication of threat to our national security; (2) there has been no evidence submitted to or discovered by the Air Force that sightings categorized as 'unidentified' represent technological developments or principles beyond the range of present day scientific knowledge; and (3) there has been no evidence indicating that sightings categorized as 'unidentified' are extraterrestrial vehicles."

On December 17, 1969, the Secretary of the Air Force announced the termination of Project Blue Book.

"There is no likelihood of renewed Air Force involvement in this area," he stated.

It was a statement, I soon realized, that was never intended to be honored.

SEVEN.

Despite all its exhaustive claims, I began to have doubts as I started grappling with the legacy of Project Blue Book. Reading through its many cartons of files, I encountered a jumble of vague, unorganized and often flippant pronouncements. "Unfortunately, I do not have interplanetary Blue Cross coverage!" complained a report submitted by an investigator after he was pricked by a thorny bush near one alleged New Mexico sighting. Too many other reports, I found, shared a similar, irritably judgmental, unscientific tone. And after a trip to its former offices at Wright Patterson Air Force Base in Dayton, Ohio, my suspicions were reinforced.

The headquarters of Project Blue Book had been lovingly preserved. It was a small, gray room cluttered with the empty desks of the project's permanent staff: a major, a sergeant, and a secretary to answer the three phones. There were no computers, no high-speed cameras, no laboratories for scientific analysis. It was more reminiscent of a college newspaper office than a sophisticated research facility. I stood there and imagined an excited call coming in, and a bored secretary fielding it: "Sure, just leave your name and address and maybe next month we'll get around to looking into it." Yet against one battleship-gray wall there was an impressive line of filing cabinets. Hundreds of drawers. Room for thousands and thousands of files. Millions of pages. And that, I decided glumly, was what Project Blue Book was really all about. Its purpose was to provide a rationale, a stream of fruitless reports that would justify the Air Force's decision to abandon its public probe of unidentified flying objects.

But, I also learned, that probe continued behind the scenes, contrary to official Air Force policy, contrary to its pledge simply to walk away from the mystery. For, when I returned East and resumed my journey through once restricted government archives, I soon unearthed proof of a more complicated history. In the period following the termination of Project Blue Book, the government's concern over UFOs had, in fact, intensified. The sightings now filled the Air Force with terror. The unidentified flying objects had become aggressive.

Throughout the fall of 1975 and the following winter, according to the pile of once classified reports I now eagerly read, the Strategic Air Command had instituted a series of Security Option Three Alerts at its northern tier bases. Unidentified flying objects were encroaching upon the Air Force's nuclear weapons supply depots.

The first incident had occurred at Loring Air Force Base in Limestone, Maine. A Flash message was sent:

"The past two evenings . . . an unidentified helicopter has been observed hovering over and in the near vicinity. Attempts to identify this aircraft have so far met with negative results . . ."

The same week as the incident in Maine, Malmstrom AFB in Great Falls, Montana, issued its own Flash:

"At 405 EST, SAC Site L-5 observed one object accelerate, and climb rapidly to a point in altitude where it became indistinguishable from the stars. NORAD will carry this incident as a FADE remaining UNKNOWN at 320 EST since after that time only visual sightings occurred."

Two weeks later, the SAC facility in Minot, North Dakota, went on alert:

"UFO sighting reported by Minot Air Force Station, a bright starlike object observed in the west, moving east, about the size of a car . . . Approximately 1120, the object passed over the radar station, 1000–2000 feet high . . . National Command Operations Center notified."

Similar sightings continued for eight months. The Joint Chiefs of Staff demanded daily updates on the "UFO Intrusions at SAC Bases." Flash messages bombarded the Pentagon. Each was an unanswered question, part of an increasingly ominous mystery. And the eyewitness testimony kept on mounting. The unidentified flying objects, according to top-secret Air Force reports, "demonstrated a clear intent in the weapons storage area"; "shots allegedly fired at aircraft at the Grand Rapids AFB alert facility"; "one individual stated the object was about the size of a 2½ ton truck."

A puzzled and concerned commander in chief of NORAD sent a confidential briefing to all units under his command. Its subject—"Suspicious Unknown Air Activity":

"Since 28 Oct 75 numerous reports of suspicious objects have been received at the NORAD COC [North American Air Defense Combat Operations Center]. Reliable military personnel at Loring AFB, Maine, Wurtsmith AFB, Michigan, Malmstrom AFB, Minot AFB, and Canadian Forces Station, Falconbridge, Ontario, Canada, have visually sighted suspicious objects.

"Objects at Loring and Wurtsmith were characterized to be helicopters. Missile site personnel, security alert teams and Air Defense personnel at Malmstrom Montana reported object which sounded like a jet aircraft. FAA advised 'There were no jet aircraft in the vicinity.' Malmstrom search and height finder radars carried the object between 9,000 ft and 15,600 ft at a speed of seven knots. . . .

"F-106s scrambled from Malmstrom could not make contact due to darkness and low altitude. Site personnel reported the objects as low as 200 ft and said that as the interceptors approached the lights went out. After the interceptors had passed the lights came on again. One hour after the F-106s returned to base, missile site personnel reported the object increased to a high speed, raised in altitude and could not be discerned from the stars. . . .

"I have expressed my concern to SAFOI [Air Force Information Office] that we come up soonest with a proposed answer to queries from the press to prevent overreaction by the public to reports by the media that may be blown out of proportion. To date efforts by Air Guard helicopters, SAC and NORAD F-106s have failed to produce positive ID."

Then, as mysteriously as they had started, the intrusions stopped. There were no more sightings. And no answers: The Air Force was never able to identify the craft that had penetrated the protected airspace above its nuclear bases. It could never explain how these craft had managed to vanish without a trace.

It was a time of growing fears. At a briefing of the Joint Chiefs, I discovered, it was theorized that these craft, so quick to vanish, were scouting ships preparing the way for an alien invasion force. It was decided, therefore, that the time had come to confront this potential enemy. The Air Force—despite all its vociferous public statements about disbelief—covertly began to prepare to do battle with extraterrestrials.

In November 1961, Air Force Intelligence, according to the reports in restricted files, decided to select personnel for two secret projects—Moon Dust and Blue Fly. A classified memo, written by a Colonel Betz, defined the objectives of these operations:

"Moon Dust: As a specialized aspect of its over-all materiel exploitation program, Headquarters USAF has established Project Moon Dust to locate, recover, and deliver descended foreign space vehicles."

And, with similar caution, Colonel Betz explained its companion mission:

"Blue Fly: Operation Blue Fly has been established to facilitate expeditious delivery to FTD [the Air Force's Foreign Technology Division based at Wright Patterson Field in Ohio] of Moon Dust or other items of great technical intelligence interest."

By the third single-spaced page, however, the colonel finally revealed the primary mission for these Moon Dust and Blue Fly teams. The "foreign space vehicles" were of extraterrestrial origin:

"These . . . peacetime projects all involve a potential for employment of qualified field intelligence personnel on a quick reaction basis to recover or perform field exploitation of unidentified flying objects. . . ."

And, finally picking up speed, the colonel proceeded to share his vision of a team possessing "the capability to gain rapid access, regardless of location, to recover or perform field exploitation, to communicate and provide intelligence reports. . . ." That is, a UFO SWAT team, a military unit trained to take someone (or something) to their leader.

At that point, with the teams poised on paper for the signal to charge into action, the Blue Fly and Moon Dust files died. Which is, no doubt, logical. How could the Air Force retrieve something unless it had landed? Still, it remained restless. The Air Force's Directorate of Operational Intelligence, according to its files, embarked in the 1960s and 1970s on its own vigilant program. It went ahead with plans for weapons to attack alien spacecraft.

USAF 7795 was the code number for one of the Air Force's earliest antisatellite weapons systems. Compared to the extravagant Star Wars designs of today, it was quite primitive. One of its design proposals introduced graffiti as a weapon in space. An ASAT would move in on an orbiting Soviet reconnaissance satellite and, when it was in position, the command would be sent from earth to spray-paint the Soviet bird's windows. But USAF 7795 also gave birth to visionary schemes—the plans for Projects Saint and Blue Gemini.

Saint—a God-is-on-our-side bit of shorthand for a more workmanlike name, Satellite Inspector—was, in reality, an orbital UFO inspector. A television camera and a radar system were to be affixed to the standard Agena B satellite already in use by the CIA. Then, after being launched by an Atlas rocket, Saint would move in on a UFO, getting to within fifty feet of the intruder. Its mission, however, was only reconnaissance. It would relay pictures to NORAD so the Air Force (and presumably the Moon Dust teams) would know what was coming.

Blue Gemini was designed as the "killer." The proposal called for a military version of the National Aeronautics and Space Administration's manned Gemini capsule to "approach, capture and disable an uncooperative satellite or unidentified flying object."

It was, I found out with some disappointment as I read on, never built. And with that last imaginative R & D scheme, another seemingly rich vein, another promising file, began to peter out. The pattern was consistent. No matter what new twist or turn I pursued through the maze of documents, the journey always ended in a dead end. For while I had worked my way through an intriguing secret library of plans, strategies, and concerns, there was still one file missing: the one with proof of flying saucers, with proof of the existence of extraterrestrial life.

And, I learned, General Pfautz's burrowers had taken a similarly frustrating trip. In his lecture in the Tank on that March morning, the general began by announcing what his team had discovered: Though the trail was complex and often provocative, the UFO mystery remained, even in classified circles, an unsolved case.

EIGHT.

Was it simply a matter of misguided enthusiasm? An error of judgment caused by the sheer joy of being back on the inside again? Or, perhaps it was more telling, the sort of professional excess that, according to his detractors, had cast a shadow on the general's career in intelligence work? Whatever the reason, as General James Pfautz continued his presentation that morning to the UFO Working Group, he, according to reports, began to ramble. First he did so cautiously; but before he was done, the general wound up overstepping his authority by leaps and bounds.

Things began to fall apart just after the general, having spoken for nearly half an hour, appeared to have reached his conclusion. "We never got our hands on the smoking gun," he complained, according to one man who was in the Tank. And, even more vividly, this witness can still recall what happened next. Because the general, his story apparently told, did not relinquish the floor. Instead, he gripped the sides of the lectern as if it were a ship's rail and there were rough seas ahead; and then, without pause, he set off in a totally unexpected direction.

The seventeen men sitting around the walnut conference table now listened as the general began to praise the DIA for its decision to establish the Working Group. A solution to the UFO mystery, he said, was vital. National security was at stake. The airspace of the United States of America must be protected against any and all invaders.

His tone was impassioned; one man in the room was convinced the general was nearly shouting. Nearly everyone had stopped taking notes. The general, though, was relentless. He was standing at a lectern in a surveillance-proof room in the subbasement of the Pentagon, but his manner, at least two of those present felt, was more suited to a revival meeting.

There was much, he told the Working Group, that the government might be able to learn if we are finally able to make contact with extraterrestrial beings. Someday, he went on, the United States might be able to obtain fantastic technology for free from an advanced and munificent alien civilization.

By this point, it was noted, Colonel Phillips was searching the other faces around the table. He was glowering; and, perhaps unconsciously, he had drawn a large *X* on the yellow pad in front of him.

General Pfautz, however, was undeterred. Imagine, he was reported as having said in a fiery conclusion, if we could enter into an alliance, a partnership with the aliens. Imagine, he said in substance, if the Air Force could get Stealth bomber technology without having to pay billions for it. And with that last shared vision of the future, he thanked the Working Group for their attention and stepped away from the lectern.

It was just as the general was apparently preparing to take a seat at the table that Colonel Phillips approached him. The Working Group operations, he abruptly explained as an uneasy silence filled the room, were compartmented.

The general didn't say anything. He stood his ground. He didn't move, but he didn't sit either. It was as if he needed time to sort through the implications of the colonel's terse statement. Sensitive Compartmented Information was a restricted classification higher than Top Secret. Certainly, it was a classification he no longer possessed. The colonel, then, was telling him he was not cleared for the rest of the proceedings. *They were telling him to leave.*

"I see," said the general at last.

And so a disappointed General Pfautz walked out of the Tank, out of the loop, and back into retirement. He could only mull over the reasons for his banishment; while behind closed and guarded doors, the Working Group would continue its exploration of forbidden provinces, secrets he would only be allowed to imagine.

Yet when I contacted General Pfautz to discuss the 1983 flying saucer task force he had initiated as head of Air Force Intelligence as well as his presentation to the UFO Working Group, he, after some elaborate coaxing, agreed to an interview at his home.

The night before our scheduled meeting he called. His tone was apologetic, but firm. "I'm afraid I'm going to have to cancel our appointment." He had spoken with "some friends at the Pentagon," he explained, and had been told that "any public discussion of the UFO Working Group would be out of the question." "I hadn't realized," he went on, "how highly classified the activities of the Working Group still are. They cannot be discussed."

Of course, I was not prepared to let the general slip away that easily. But, I realized, it was unlikely that any of my thin arguments would prove very persuasive that night. Or for a while. I decided to let a calming week or two pass before making another attempt.

I did not need to wait the week. Just three days later the general called again. He seemed, at least to my ear, to be in a state of high nervousness. "How could you do that?" he demanded.

"What?" I asked with as much innocence as I could command. And yet I was genuinely wondering what I could have done.

"How could you have said that to General Stubblebine? Where do you get off telling him I talked to you about the 1983 study? About the Working Group?"

"I didn't," I insisted with some indignation. And, carefully, taking pains to re-create the conversation as precisely as possible without my notes in front of me, I shared a summary of my brief interview earlier that day with retired Major General Albert Stubblebine, a former head of Army Intelligence. I had asked General Stubblebine about his role in the joint Air Force-Army-DIA flying saucer task force that General Pfautz had organized. And I had also questioned him about his knowledge of the UFO Working Group. Was he present, I had asked, when General Pfautz had made his speech?

General Stubblebine insisted he had no knowledge of either the task force or the Working Group. And with that, the interview ended.

But not the episode. Someone had obviously passed on to General Pfautz a maliciously twisted version of my conversation with General Stubblebine. Pfautz was enraged: "I have never revealed classified information and I would never do so now. How could you let them think I had?" He suggested that in the future I get all my information from the Public Affairs Office of the Department of Defense. And with a curt "Good-bye, sir," he then broke the connection.

Later that week I wrote him a note. I pointed out that according to the Department of Defense there was neither a 1983 flying saucer task force nor a UFO Working Group. I concluded, "We both know that is simply not true."

And, I might have added, I had proof. Even without the general's help, by now I knew too much.

Part III
THE ORDER OF THE DOLPHIN

NINE.

It was waiting for them when they entered the Tank. As the members of the UFO Working Group took their places at the long conference table on a spring morning in 1987, they found that a two-page briefing paper had been centered in front of each seat. It was stamped "Secret" and, according to the letterhead at the top of the first page, it had been prepared by the President's Office of Science and Technology.

This document also fell into my hands. It began:

"Over the past two decades there has developed an increasingly serious debate about the existence of extraterrestrial intelligent life. Scientists working for the National Aeronautics and Space Administration are convinced (1) that intelligent extraterrestrial life exists and (2) that NASA has developed the only rational way for detecting advanced intergalactic civilizations. The program, called SETI for the Search for Extraterrestrial Intelligence, will cost over $97 million and involves the use of a network of radio telescopes to detect signals sent from other worlds. . . ."

And I learned that it was this secret briefing paper that had convinced Colonel Phillips the time had come for the Working Group to launch its first field operation. Since the inquiry would involve a knowledge of signal detection procedures and electronics, he recommended that the operatives be "borrowed" from the National Security Agency. The mission's purpose: to determine if the government of the United States of America could make contact with other worlds through the SETI network.

But there was, I later found out, another reason why Colonel Phillips wanted to recruit his first field team from the NSA. He was hoping for agents who would be sympathetic to the possibility of extraterrestrial life, and he knew he would find believers in the Dundee Society, an elite, almost fraternal group of career officers within the NSA. It functioned as a secret society within a secret—arguably the nation's most secret—organization. And it also was a group that from its inception had proselytized within the NSA that flying saucers had, most probably, already visited Earth.

The founder of the Dundee Society was Lambros D. Callimahos, flutist, linguist, cryptologist, the author of perhaps more classified monographs and books than any other nonnuclear thinker, and generally a man who was, thanks in no small part to his own elaborate efforts, the stuff of legends. He would preside over meetings of the Dundee Society resplendent in a fitted beige Nehru jacket, white trousers woven, or so went the giddy scuttlebutt, from the hair of virgin llamas, and matching white shoes crafted (more legend?) by reformed anthropophagi from the Andaman Islands. On his desk was enshrined the totem that had inspired the society's name—a Dundee marmalade crock filled with four well-sharpened pencils; the world's first code-breaking machine, he had called it. And, more ritual, the Dundee Society initiates lavished on him an assortment of deferential titles—The Nameless One, His Cerebral Phosphorescence, The Guru, and The Caudillo.

The meetings of the Dundee Society were free-wheeling discussion groups, the subjects ranging from cryptology to Kantian philosophy to animal rights. One of the most frequent topics for debate, however, was inspired by a classified paper the society's founder had written.

Entitled "UFO Hypothesis and Survival Questions," and written for restricted intra-agency distribution nine years before his death in 1977, this seven-page Socratic discussion of the issue was, by shrewd design, determinedly argumentative. The product of a teacher's mindset, the monograph raised succinct hypothetical questions and then, obviously relishing the challenge, had some fun answering them. In many ways, it echoed the spirit of a typical Dundee Society meeting.

The monograph began by considering the hypothesis "all UFOs are hoaxes." And it concluded: "The fact that UFO phenomena have been witnessed all over the world from ancient times and by considerable numbers of reputable scientists in recent times indicates rather strongly that UFOs are not all hoaxes." Then, wondering if perhaps "all UFOs are hallucinations," The Nameless One decided: ". . . sometimes a person and a gun camera confirm each other's testimony. On occasion, physical evidence of a circumstantial nature was reported to have been found to support witnessed sightings. A continuing high percentage of reports of unusual aerial objects are being reported by people in responsible positions in science, government, and industry. The sum of such evidence seems to argue strongly against all UFOs being hallucinations."

The paper then moved on to ponder the possibility that "all UFOs are natural phenomena." Here a warning was rumbled: "Many UFOs have been reported by trained military observers to behave like high speed, high performance, high altitude rockets or aircrafts. . . . Sometimes the phenomena appear to defy radar detection and to cause massive electromagnetic interference. Surely it is very important to discover the nature of these objects or plasmas before any prospective enemy can use their properties to build a device or system to circumvent or jam our air and space detection systems. . . ." Which led to a final question: Are UFOs the products of extraterrestrial intelligence? An equivocal position was stated unequivocally: "According to some eminent scientists . . . this hypothesis cannot be disregarded."

But it was in its chatty concluding section, where a parable of sorts was offered, that the monograph was to have its most significant effect on subsequent NSA activities:

"If you are walking along a forest path and someone yells 'rattler,' your reaction would be immediate and defensive. You would not take time to speculate before you act. You would have to treat the alarm as if it were a real and immediate threat to your survival. Investigation would become an intensive emergency action to isolate the threat and to determine its precise nature. It would be geared to developing adequate defensive measures in a minimum amount of time. It would seem a little more of this survival attitude is called for in dealing with the UFO problem."

With the telling and retelling of this parable over the next decade, an operational way of looking at the world evolved at the NSA—one that divided the agency into two contentious factions. On one side were the Dundee Society initiates and their starry-eyed protégés. These NSA officials, by inclination and profession ponderers, treated the wisdom of their Guru and Caudillo with a hermeneutical deference; and in his scripture what had become known throughout the agency as "The Tale of the Rattler" loomed large. Its moral: Shoot first—and ask questions later. Or, in strategic terms, a perceived threat, whether it be from a foreign or even an extraterrestrial nation, should be dealt with as if it were real; there will be time afterward to debate its authenticity.

Such a defensive way of thinking, however, had its detractors. "Sick think," they called it. This camp contended an intelligence organization has only so many resources at its disposal; to squander time, men, and money on nonexistent threats was a dangerous indulgence. Giving the Guru's analogy their own twist, they argued that it doesn't always make good sense to leap if someone yells "rattler." Imagine you're strolling along the sidewalks of New York; only a fool—or a paranoid—would jump. And some troublemakers even took the argument further. They nastily suggested that any mind obsessed with snakes was a tangled one to begin with; undoubtedly it fretted over little green men, too.

But in the end, the Dundee Society won out. The National Security Agency, contrary to all its public statements, has since 1972 been secretly monitoring and often assessing worldwide allegations of UFO activity. For the past eighteen years, NSA listening posts throughout the world have operated under a standing order. It is mandatory to Flash-report Fort Meade on all intercepted foreign-government signals intelligence that pertains to unidentified flying objects; and these installations are required to track and Flash-report on any signals or electronics intelligence that might have an extraterrestrial origin.

And so in the week before Memorial Day, 1987, a team was recruited by the UFO Working Group from the blackboard-lined offices in the NSA's Research and Engineering Division at Fort Meade. The men were veterans of the Intercept Equipment Division, and they all were members of the Dundee Society. They reported to the Pentagon office of the DIA to receive their formal instructions. The orders had previously been drafted by Colonel Phillips and approved by the entire Working Group. They were:

". . . to examine systematically the validity of the fundamental criteria and axiom associated with the National Aeronautics and Space Administrations's SETI program to detect extraterrestrial intelligent life; and to estimate the current probability of success of the SETI program. . . ."

The next day when a NASA jet made its approach for a landing at a restricted military base in the Mojave Desert, on board were two NSA operatives on assignment from the UFO Working Group.

Eighteen months later, I set out to follow in their footsteps. At just about dawn, I arrived at the Tiger Air Terminal in a remote corner of the Burbank, California, Airport. The lights in the terminal were still dark, but—and this came as a bit of a surprise—there was an attentive woman with red lips as bright as a beacon behind the front desk. She was wearing a midnight blue outfit that despite its lack of insignia was meant to suggest a uniform, and as soon as anyone entered the dim terminal, she announced her presence in a cheery warble. "Please check in," she called out with polite assertiveness. "Up here, please."

The check-in procedure was very simple. There were no tickets for this early morning flight. Either your name was on the list, or it wasn't. Two documents confirming your identity were also required. The woman in the blue uniform considered each of the offered documents with the conscientious attention of a jeweler scrutinizing a gem. And while she did this, I noticed something else. There was a holster at her narrow waist holding a small black revolver.

I was cleared without any problems. And after the time it took to finish a cup of black coffee, I was allowed on board an eight-seater NASA jet. My fellow travelers, I decided, fell into two groups: A few were military types traveling out of mufti, but their graying crew cuts and joggers' physiques were as identifying as any uniform. The others were men in windbreakers and open-necked shirts, youngish and enthusiastic, looking at first glance like duffers getting away for a quick eighteen holes. But the row of pens, mechanical pencils, and ultrathin pocket calculators lined up in their breast pockets gave them away. They were the engineers, the R & D whiz kids flying off on assignment to tighten screws or simply, as proud parents, grabbing a chance to watch their latest creations in action.

Suddenly—there was no announcement—the small plane was rising high above the early morning Southern California smog. It was heading due northeast, toward the Mojave Desert. I was on my way to NASA's goldstone complex in the Fort Irwin Military Reservation in Barstow, California, where I, too, would try to learn if the government had developed a way to communicate with other worlds.

TEN.

SETI began, appropriately, with music. The inspiration for NASA's ninety-seven-million-dollar program to search for extraterrestrial intelligence struck as Bach was being played. It was at Cornell University in the spring of 1959 while a chamber music quartet was sending Bach's sweet message across time and space that Phillip Morrison had his idea. He was seated near the rear of the student center, his eyes closed, his head leaning toward the stage. The majesty of the music was exciting. And Professor Morrison's mind was spurred on by it, coaxed by it, until, inspired, his thoughts started wandering to somewhere else.

The first musing seemed to sneak up on him, taking him by surprise. Are we alone in the universe? He was almost startled by the suddenness of his mind's unexpected question; but, without anyone else's even noticing the first effects of the brisk intellectual adventure about to unfold, he quickly recovered. Of course, he was by that stage in his life accustomed to these flashes of problem-solving. He was a forty-four-year-old physicist, a former group leader at the Los Alamos atom bomb project. He enjoyed and coveted his mind's rambles. Besides, the quartet was a soothing and fertile background. So he took a step backward.

Why am I at this moment thinking about extraterrestrials? he asked himself. Something someone said in the lab? No. Something I read? Yes, that must be it. And his mind quickly reviewed a small stack of scientific journals until he settled on an unlikely source. His inspiration had come from the book on his night table—the astronomy volume of Joseph Needham's *Science and Civilisation in China*. It was a literate and lucid history, a work more of philosophy than pure science, and Morrison, a voracious reader, had been galloping through it simply for fun. And now as he was seated in the student center, Bach surrounding him, the gist of some lines he had read came back to him with the resonance of an irrefutable wisdom. It was (and he would locate the exact words later, on page 221 of volume 3) an insight by the thirteenth-century philosopher Teng Mu:

"Upon one tree there are many fruits, and in one kingdom many people. How unreasonable it would be to suppose that besides the heaven and Earth which we can see, there are no other heavens and no other Earths?"

How unreasonable indeed! It made no sense, instinctive or rational, Morrison silently agreed, to assume that we on Earth are alone in the universe. But if others are out there, how can we reach to them?

This new question rumbled about his mind as the concert continued. He toyed with it for a while. The intensity of his concentration was remarkable; he might just as well have been alone in the student center. And then, his pace quickening, the pieces finally beginning to link in a neat, logical chain, his mind took another giant step backward.

He recalled a conversation in his office a week or so earlier. It all came back to him very clearly. He had been sitting at his desk, staring out over Cayuga Lake toward the green hills of Ithaca. There had been a knock on his door and his colleague Giuseppe Cocconi entered. Professor Cocconi, always passionate about new ideas, had come to discuss Morrison's recent paper on theoretical gamma-ray astronomy. Gamma rays, as the paper had pointed out, are electromagnetically produced short-wavelength radiation similar to X rays. That afternoon Cocconi, challenged by Morrison's speculations about the possible uses for gamma rays, had come to suggest one of his own.

There are few natural emitters of gamma rays in the sky, Cocconi began. Morrison agreed, so the physicist forged ahead. Therefore, if we observe an unexpected source of gamma rays, it would undoubtedly be worth examining. In fact—and Cocconi, smiling, teasing, playing, built to his deduction with some fanfare—these "unnatural" gamma rays might very well be a beacon—*an attempt by one planet to signal another.*

At the time, Morrison now recalled from his seat in the crowded student center, he had been intrigued, enthusiastic—and then dismissive. The concept of a planet's signaling across space to Earth was inspirational. Why shouldn't planets try to communicate with each other? And Cocconi's strategy for a passive search rather than the sending of signals made eminently good sense. It wasn't simply that no one knew where to aim a beacon, it was also good interstellar manners not to go shining bright lights in the eyes of any possibly bellicose extraterrestrials. But the more Morrison had poked at the concept, the more it seemed to deflate. Gamma rays, he had explained to Cocconi, were difficult to receive as well as generate. A synchrotron powerful enough to produce a concentrated beam of gamma rays would cost millions of dollars—if it could even be built. No, he had finally told his colleague, it was an intriguing idea, all right, to scan space for a gamma-ray signal from another civilization. But it just wasn't practical.

And so the suggestion, and the conversation, had been discarded into a cluttered corner of Morrison's rich brain. Until, with the Bach being played in the background, the words from Teng Mu providing further provocation, it was all recalled and this time, in the Cornell Student Center, the ideas bubbled and swirled in the professor's mind until they came together with the force of one clear, penetrating insight: *Radio waves*. A low-frequency radio wave, like a high-frequency gamma ray at the other end of the electromagnetic spectrum, traveled at the speed of light; radio signals could also reach across the universe to Earth from distances of many thousands of light-years. But, unlike gamma rays, radio waves were easy to transmit and receive. A short message could be sent between two interplanetary receiving stations twelve light-years apart at the cost of one New York City subway token. And rapidly, his mind racing, the structure of an experiment began to form. Why not use already existing radio telescopes to listen for electromagnetic signals—radio or television transmissions—beamed into space by alien civilizations? It was the only rational way.

Suddenly, there was applause. Can they read my mind? Morrison might very well have thought. But onstage the four musicians were taking bows, and the professor was abruptly pulled back to the everyday. Nevertheless, Phillip Morrison left the student center confident that in a random hour's musing he had hit upon an idea, a strategy for communicating across space, that one day would change life on Earth forever.

ELEVEN.

Phillip Morrison was not the first man on Earth to imagine communicating with other worlds. His idea was just the most recent in a long historical procession of schemes and dreams. But, NASA would decide, it was the one plan with the greatest chance of success.

Not that there had been much competition. Earlier attempts to get in touch with extraterrestrial civilizations had been wonderfully inventive—and wonderfully wrong-headed. There was Carl Friedrich Gauss's, for example. In the 1820s, this German mathematician announced he had come up with the perfect way to alert the inhabitants of the Moon to the existence of earthlings. He wanted to plant groves of pine trees across Siberia—hundreds of acres of trees laid out in the shape of squares on the sides of a right triangle. This design—a Victorian earthwork—would be a signal. Any man on the Moon peering through his observatory telescope would be able to see there was life on Earth sufficiently intelligent to have mastered the Pythagorean theorem. The sum of the square of the number of pine trees that formed the legs of the vast right triangle would be equal to the number of trees in the square of its long hypotenuse.

About twenty years later Joseph von Littrow, an astronomer, had what he was convinced was a better idea. His plan was to dig a twenty-mile ditch across the Sahara Desert, fill it with kerosene, and set it ablaze. In the dark night sky, this fire would serve as a beacon across the universe.

The trees were never planted. The ditch was never dug. None of those primitive SETI projects was ever built. Still, like so many wishful thoughts, new proposals for communicating with extraterrestrials continued to be suggested by visionary thinkers. There was even one for building a highway of reflecting mirrors across Europe. It, too, was ignored. It wasn't until Percival Lowell "discovered" men on Mars that a largely private obsession took a firm grip on the American consciousness.

Lowell was a regal, waistcoated Boston Brahmin who flatly rejected the fussy restraints of his mercantile heritage and, instead, looked for stronger solace and inspiration in the beauty and mystery of the heavens. Confidently, he walked away from Boston and headed into the Arizona desert with its wide, open sky. In 1893 he built the Lowell Observatory in what was then the remote town of Flagstaff. It was there that Lowell busied himself with the study of Mars—and, he was convinced, its inhabitants.

Lowell, a model of dedication, stared into his telescope until, triumphant, he determined that the dark shapes he was observing had to be nothing less than a planetwide network of deep "canals" across Mars; and these canals, he went on to deduce, carried water to a vast system of sophisticated Martian farms. More significantly, he shared his excited discoveries with the public: His books *Mars and Its Canals* (1906) and *Mars as the Abode of Life* (1908) were best-sellers. All across America people were looking up at the sky, now certain someone was looking back at them.

It became a national craze. As America entered the twentieth century, there were newspaper contests ("Tell Us Your Idea for Talking to Mars") and even songs about "the lonely man on Mars." Armed with the new century's fledgling technology, scientists and laymen tried to solve the riddle. Robert Wood, an American physicist, had the idea of positioning large black sheets across miles of white drop cloth; motors would undulate the sheets, and so, from a celestial vantage point, a flirtatious Earth would appear to be "winking" provocatively at Mars. A prescient British scientist, Sir Francis Galton, invented a code based on Morse that would transmit the outlines of a picture. And, a variation on an old idea, it was suggested that an array of mirrors flash out a slow-speed semaphore message to Mars.

But by 1942, when the first successful V-2 rocket was launched by German scientists at Peenemünde on the Baltic Sea, these schemes had become the butt of jokes. The future was certain. Contact with extraterrestrial life would happen through space travel. Spaceships would carry explorers to other worlds.

It would take only another generation for many physicists to decide that this, too, was wishful thinking. They had become convinced that manned space travel beyond the solar system made as much sense as winking at the heavens with black flags.

. . .

Space was too vast. The stars were too far away. Time, physics, and economics all conspired to make manned space travel across the galaxies improbable. Proxima Centauri, the nearest star to the Sun, was four and a half light-years away. Andromeda, the nearest galaxy similar to the Milky Way, was about two million light-years from Earth. How were astronauts to travel these distances?

Consider the fastest space vehicle mankind has ever produced, Pioneer 10. Its payload stage traveled at 25 miles per second. But even at this speed, it would take Pioneer 10 some 33,000 years to reach Proxima Centauri. A trip across the Milky Way to the Andromeda galaxy in Pioneer 10 would take about 15 billion years—which is as long as the universe (and time) have existed.

But if chemical-fueled rockets were too slow, perhaps science could create another fuel system? Perhaps something could be invented that would push a manned vehicle to warp drive speeds like a ship out of "Star Trek"? The most effective drive system physicists can imagine would be powered by antimatter. This design, if it were ever built, would convert nearly all the fuel's energy to mass, and—kaboom!—the radiation from this annihilation would be the rocket's exhaust.

So, how fast could this antimatter machine travel? According to Albert Einstein, there is a celestial speed limit. Nothing can travel faster than the speed of light. No vehicle—not even a ship powered by an antimatter drive system—could *ever* travel faster than 186,000 miles per second. Even if physicists managed to master the science necessary for this fuel system, voyages across the galaxies would still stretch on for decades.

And these ships would require enormous amounts of fuel. Another of Einstein's axioms established that the faster an object moves, the greater its mass. If a spacecraft were traveling at near the speed of light, then its mass would be approaching infinity. A vehicle that heavy would need to be propelled by vast amounts of energy.

An example: An antimatter ship is preparing to embark on a ten-year trip around the heavens. It will travel at seven-tenths the speed of light. By the end of its journey, therefore, it will have annihilated 33,000 tons of antimatter fuel. The total energy released would have been enough to supply the entire present electrical power needs of the United States for the next half-million years.

Then there would be the cost. The design and construction of the spaceship would run into billions of dollars. But, when compared to the fuel bill, that would be small change. Antimatter can presently be produced in only a few subatomic particles at a time. Today it would be priceless. In the future, in the twenty-first century perhaps, scientists predict they will be able to manufacture antimatter. At a price—ten million dollars a milligram. Which means a ton would cost nine quadrillion dollars. And, of course, the starship making the ten-year voyage would need 33,000 tons.

The conclusion was unavoidable. Fantastic voyages of manned spaceships to planets beyond the solar system would occur only in science fiction. Science had to find another way to communicate with other worlds.

Phillip Morrison agreed. Long before the momentous chamber music recital, he had, much to his intellectual disappointment, already reached the sour conclusion that men were never going to travel to Andromeda. The obstacles, he had decided with flat objectivity, were insurmountable. And so his thoughts had turned to radio waves as a means of communicating across vast space. It was, he repeatedly told himself, the only rational way. But then in the aftermath of inspiration, Professor Morrison came to realize there was at least one immediate and looming problem with his theory, too. He knew next to nothing about the field of radio astronomy.

TWELVE.

As the spring term moved toward finals, Morrison and Cocconi, like most of the students at Cornell, were busy cramming. Before they could present their plan—and it was a joint effort—to their fellow scientists, they decided they needed to learn more about radio technique.

Not that there was much literature in the field. Radio astronomy was only a generation old, the product of an accidental discovery by Karl G. Jansky. In 1931, Jansky, a radio engineer at the Bell Telephone Laboratories, was attempting to track down the sources of the annoying high-frequency static which had been interrupting transoceanic phone calls. As part of his research, Jansky, a methodical as well as a resourceful engineer, had built a device he called his "Merry-Go-Round." It was a crude, makeshift structure. A row of tall, thin rectangular antennas were lined like open doorways on a hundred-foot-long wooden frame; the frame rode on four wheels salvaged from a Model T Ford and a small motor rotated this jittery contraption every twenty minutes. Still, the Merry-Go-Round was effective. Jansky quickly began to zero in on what was causing the static.

One source, he was certain, was lightning from nearby thunderstorms. Also distant storms, he discovered, would wreak havoc with communications; their radio emissions were bent back to Earth by the ionized regions of the upper air. But then he encountered something that was more puzzling. A loud hiss was coming from the sky no matter where his Merry-Go-Round's antennas were aimed. At first he thought the noise was an effect of sunlight on the Earth's atmosphere, but he soon abandoned this theory. Instead, at a meeting of the American Section of the International Scientific Radio Union on April 27, 1933, he announced his conclusion. The radio emissions he was receiving came from the midst of the solar system. The heavens were, he proclaimed, an immense, noisy place. The natural activity of the stars, nebulae, and galaxies created a cosmic static.

Jansky's discovery was front-page news in *The New York Times*. Professional astronomers, however, were largely uninterested. In fact, the only person who seemed to act on Jansky's findings was a ham radio hobbyist. In his small backyard in suburban Wheaton, Illinois, Grote Reber confounded his neighbors by erecting a parabolic antenna thirty-one feet in diameter. He used this big "dinner plate" (as the puzzled neighbors called it), a forerunner of the modern radio telescope, to eavesdrop on the sounds of the universe. For the next decade, he was the world's only active radio astronomer. His white-picket-fenced yard was the only listening post on Earth tuned into the world out there.

By the time Morrison and Cocconi were refining their plan nearly twenty years later, radio astronomy had become an active field. A radio telescope was simply a big antenna with a receiver attached; the antenna collected radio waves from as broad an area as possible and focused these electromagnetic signals so that they could be conveyed to the receiver. But, astronomers were learning, it could be an effective instrument. Radio waves from space had revealed eruptions on the Sun, the temperature of the Moon and nearby planets, and the presence of frantically gyrating atomic particles in the radiation belts surrounding Earth and Jupiter, as well as in the stormy gaseous clouds of the Crab Nebula.

And, the two Cornell professors discovered, mammoth instruments, radio telescopes that would have towered over Reber's entire house, had already been built. One, at Jodrell Bank in England, had an 800-ton receiving dish that was 250 feet long, a length only a bit smaller than a football field. Another was in Green Bank, West Virginia, where the National Radio Astronomy Observatory's newly constructed 85-foot dish rose menacingly above a lush valley surrounded by the Blue Ridge Mountains.

Morrison and Cocconi were elated. The receiving and transmitting equipment required to communicate with other worlds was already in place and working. But as the two professors pressed on with their study of radio, they realized that there was another obstacle. It was a problem that would become known as the "Cosmic Haystack."

How could one "tune into" an extraterrestrial broadcast? How, Morrison and Cocconi wondered, could a scientist decide what frequency among the millions of possible frequencies in the heavens was the correct one to listen to? It was, Morrison would suggest to me in a conversation nearly three decades later in his office at MIT, like trying to find one specific radio station, the mellow, out-of-this-world sounds of W-ET—only the radio had an almost infinite number of stations on its dial. Or, another daunting way of putting it, it would be like looking for a needle in a cosmic haystack.

As soon as they began to consider the problem, they were overwhelmed by it. Cocconi, stymied, moaned, "We might as well try picking the winning numbers in the Irish Sweepstakes." Yet, oddly, that stray complaint energized his partner to continue through this rough period. It served, Morrison gaily believed, to scale down the odds. Success was, of course, possible; after all, people *do* win the Irish Sweeps.

Not everyone, though, was as optimistic as Morrison. His colleagues seemed to take some delight in picking apart his theory. It was in the midst of one of these late afternoon critiques in Morrison's office that a colleague—his name long since forgotten—argued that the experiment was impossible. After all, he sneered, the aliens could hide their signal on any frequency they chose.

But, Morrison immediately lashed back, they're not trying to hide anything. They're sending us a signal; they want us to find it. And once he said it, Morrison realized with a pounding shock of recognition that there just had to be a frequency out there that made sense: a *logical* channel. A channel that both the transmitting civilization and the receiving civilization would both instinctively understand was the single logical choice—without the luxury of prior conversation.

So, once more encouraged, Cocconi and Morrison threw themselves back into the problem. Only now they tried to think like aliens.

Suppose, they began to reason, you were station manager of W-ET, and you wanted your station to reach across the universe. Where on the noisy electromagnetic dial would you find the frequency that would give you the clearest, cleanest sound, a static-free crispness that would allow W-ET to broadcast loud and clear across the heavens?

First off, they eliminated the bands of frequencies on the low end of the dial. This part of the electromagnetic spectrum was already cluttered with thick cosmic static—star noise, dust cloud emissions, cosmic background radiation. An alien station manager would have a hard time building a transmitter that could shout above that racket.

Next, the two Cornell professors considered the frequencies on the other end of the dial. These, too, wouldn't work; the high-frequency bands were absorbed by the Earth's atmosphere.

And so by process of elimination, Cocconi and Morrison began to focus their attention on a relatively quiet stretch of the dial—the microwave region. A transmitter pitched at this frequency (anywhere from one to ten gigahertz on the electromagnetic dial) would require very little energy to shout above the noise in the universe. It, they reasoned, would be the frequency of choice of any rational alien.

But this insight left the two professors with another dilemma: Where in this microwave window, as it was called, should someone on Earth set his receiver? Pointing a radio telescope at a corner of the sky and searching for a beacon—say a monochromatic signal like those used in TV or FM broadcasting with a 1Hz bandwidth—in the microwave region would still not guarantee they would pick up W-ET. If they tuned into each microwave frequency for just ten seconds—one-third the time of most radio commercials—checking out a single direction of the sky would take more than three thousand years. They still, the professors glumly realized, needed an intergalactic radio program guide, something that would tell them precisely where on the microwave dial they could tune into station W-ET.

Phillip Morrison found the answer on a trip to Moscow.

The precise circumstances of the discovery of what has become known as "the magic frequency" have faded a bit from the memory of most of the participants. When in preparation for my trip to the Mojave, I interviewed Professor Morrison, he tried to joke about this lapse. "I've been around forever. I've even met Einstein twice. I go back that far. How should I remember what went on in Moscow?" But he was a cordial man, with easy charm, and if an inquisitor was persistent enough, he would try to cooperate.

Yes, he said, he and Cocconi had gone to Moscow to attend an international conference on cosmic rays. And yes, while there he had shared his problem with a Harvard astronomer. Was it at the end of a long day? The two scientists seated at a table in an Intourist hotel bar, a bottle of vodka and some fresh black bread handy to fuel a rambling conversation? A conversation that, as luck and alcohol would have it, just happened to turn convivially to, of all things, the search for a natural interstellar wavelength? "Not really," said Morrison quickly. But after a moment, he was more agreeable. "Oh, I don't know, maybe it was like that. Sure."

Whatever the setting, there was no question about the scientific "news" the Harvard man shared with Morrison in Moscow. Astronomers had recently determined that hydrogen, the most abundant element in the universe, emitted energy at a characteristic frequency of 1,420,405,752 times a second, which also could be expressed as 1,420 megahertz. And that was a band in the nearly noise-free microwave window.

This information was the missing link Morrison needed. An interstellar landmark, a logical meeting place, a "magic frequency" was now apparent on the microwave dial. For, as Morrison eagerly explained to Cocconi, an advanced extraterrestrial civilization, one smart enough to communicate across the galaxies with radio technology, would certainly also have detected, like us primitive earthlings, the hydrogen signal. Therefore, if these advanced extraterrestrials were hoping to attract attention to themselves, they would undoubtedly decide to broadcast their signal at or around 1,420 megahertz.

At last, the experiment—arguably the most important piece of exploration in the history of the world—was doable. It was possible, the two Cornell physicists believed, to aim a radio telescope tuned to about 1,420 megahertz (21 centimeters) into space, and maybe, just maybe, pick up a broadcast from another world.

Only now they had to convince their fellow scientists to do it.

So they began another search. This time Morrison and Cocconi were looking, they would say, for intelligent life on Earth. It took some looking.

First, Cocconi tried. On leave from Cornell to do research at CERN, the international atomic study center in Switzerland, he decided to start with the one man who had precisely what they needed—the world's largest radio telescope. He wrote to Sir Bernard Lovell, founder and director of the Jodrell Bank observatory.

"My name is probably unknown to you," he started out deferentially, and continued in this careful, measured tone. He even went so far as to acknowledge that the program he was about to suggest "at first sight . . . looks like science fiction." Still, when he got into the heart of the matter, he offered his theory cogently and boldly:

"There is a good chance that among the, say, 100 stars closest to the sun, some have planets bearing life well advanced in evolution . . . I will assume that 'beings' on these planets are already sending the stars closest to them beams of electromagnetic waves modulated in a rational way, e.g. in trains corresponding to the prime numbers. . . ."

Less than three weeks later, Sir Bernard responded. He was not impressed. Such a search, he wrote in his terse reply, would be "difficult" if not "frivolous." Years later he would regret his indifference; he would even publish Cocconi's letter as an appendix to one of his books.

It was a blow. Morrison and Cocconi, however, refused to give up. Instead, disappointment solidifying into anger, they regrouped. This time, they decided to write an article together. Perhaps, they reasoned, a scholarly journal might better appreciate the importance of what they were proposing. It took them three months. They wrote. They argued. And then they rewrote. When they were done, they gave the short piece a concise but still dramatic title—"Searching for Interstellar Communications."

And this time, learning from past mistakes, they operated according to a more pragmatic strategy. There would be no you-don't-know-me-but letters to formidable editors of prestigious journals. Instead, Morrison wrote to an old friend, the influential physicist P. M. S. Blackett, then at Imperial College, London. Would Professor Blackett, Morrison asked with some trepidation, pass on the enclosed article to the editors of *Nature*?

This, too, was a magic frequency. On September 19,1959, on pages 844 to 846 of *Nature*, sandwiched between an article on the electronic prediction of swarming in bees and one on metabolic changes induced in erythrocytes by X rays, "Searching for Interstellar Communications" appeared.

In its opening paragraph, the two physicists announced confidently: ". . . near some star rather like the Sun there are civilizations with scientific interests and with technical possibilities much greater than those now available to us . . . We shall assume that long ago they established a channel of communication that would one day become known to us, and they look forward patiently to the answering signals from the Sun which would make known to them that a new society has entered the community of intelligence."

This "channel," the article continued, was radio, specifically transmissions made over "a unique, objective standard frequency, which must be known to every observer in the universe: the outstanding radio emission line at 1420 Mc/sec of neutral hydrogen."

They even suggested suitable targets: "The first effort should be devoted to examining the closest likely stars. Among the stars within 15 light years, seven have luminosity and lifetime similar to those of our Sun. . . ." And their conclusion was a rallying cry, a call demanding an immediate commitment from the international community of science to search the heavens for other worlds:

"We submit . . . that the foregoing line of argument demonstrates that the presence of interstellar signals is entirely consistent with all we now know, and that if signals are present the means of detecting them is now at hand. Few will deny the profound importance, practical and philosophical, which the detection of interstellar communications would have. We therefore feel that a discriminating search for signals deserves a considerable effort. The probability of success is difficult to estimate; but if we never search the chance of success is zero."

Morrison was on a year's sabbatical when the article appeared, but a few keen journalists managed to track him down in remote corners of Europe and Asia to discuss his theory. He was surprised, yet delighted, at the attention the brief piece was attracting. But all this was only a small preamble. Upon his return to America, an astonished Phillip Morrison discovered something that left him filled with hope and even awe—the search had already begun.

THIRTEEN.

It is one of the commonplace ironies of science that great minds are often, unknown to each other, chasing after the same tantalizing idea. As fate would have it, at just about the time Morrison and Cocconi were submitting their article to *Nature*, a young astronomer was thinking along *precisely* the same wavelength—21 centimeters. And, while lunching in a roadside diner in the backwoods of West Virginia on a bleak, snowy day at the tail end of the cold winter of 1959, he, too, had decided the time had come to act on his thoughts.

They called the diner "Antoine's," but that was really a running gag. A sarcastic scientist, part of the small crew working nearby at the newly built National Radio Astronomy Observatory in Green Bank, had christened the place after the first gruesome communal lunch. The toney, four-starish name stuck. Yet, as that first, long winter dragged on, after months of ghastly meals, the joke was becoming difficult to sustain.

It wasn't just the diner. In those first months, life at Green Bank, as in most of rural West Virginia, was pretty hardscrabble. The weather, too, was another complication. The cold had been brutal, and then a brief thaw, once the hope of hopes of many at the observatory, had perversely brought little joy. Swamps of thick, primeval mud covered the land. As the observatory's scientists drove the five miles to "Antoine's" one eventful noon for lunch, they found themselves cheering the fresh falling snow; once dreaded, it now seemed cleansing.

Still, despite the isolation and the hardship conditions, working at the observatory was a pretty exciting tour of duty for a radio astronomer. Green Bank had a federal charter, as well as the money to back it up, to build the best radio observatory in the world. Already an eighty-five-foot telescope had been erected in a lonely, quiet valley, and construction had just begun on another larger instrument. If the opportunity to play with such wonderful toys required certain personal sacrifices, many radio astronomers were quite willing to pledge themselves to lives of relative exile, to trudge through mud, to shiver in blustery cold. Frank Drake, for one, jumped at the chance.

He had been recruited at twenty-nine, fresh out of Harvard, where he had received his doctorate in astronomy. But, as Drake himself would point out with a self-effacing stammer, the purposeful course of his life had been carved out decades earlier. He was nearly eight, an age when other boys growing up in Chicago were starting to follow the Cubs, when he first began to wonder about the universe. Were other people somewhere out there? The question popped into his mind for reasons he never understood, but once there it took root. All his education—Cornell; then Harvard; even the three years in the Navy aboard the heavy cruiser USS *Albany*, days in the radio shack, nights on deck contemplating the shining secrets of the Milky Way—was preparation for a professional life that would be devoted to gratifying a child's moment of wonder. It was logical, almost inevitable, that Drake wound up as part of the initial crew at Green Bank.

Throughout that first winter, he complained about the oppressive weather, groused about the inedible food, ranted against the colossal absurdity of trying to run a world-class observatory out of a dilapidated farmhouse, but all the while he was also having the most exhilarating experience of his life. The eighty-five-foot telescope had given him a window on the heavens. And, Drake learned, there was an unexpected bonus to life in lonesome, rural West Virginia—the time to think deeply.

On that snowy noon at "Antoine's," as the scientists sitting in the booth gobbled greasy hamburgers and even greasier french fries, Drake shared the idea he had been slowly piecing together over the past solitary months. If an artist had chosen to commemorate the moment, it would no doubt have been Norman Rockwell. For Drake was a perfect Rockwell model. Drake possessed sharp though friendly features, pure Americana in their routine handsomeness, and the sort of broad, high forehead that has always been iconographic fodder for deep thoughts and lofty values. One can see young Drake leaning across a Formica table like the polite son at a Rockwell Thanksgiving feast, an intensely shy man speaking hesitantly, yet on this occasion with the strength of utter conviction.

He began by sharing some calculations he had made. He had been wondering—and he announced this as if it were nothing more than the most natural bit of idle contemplation—just how far into space the observatory's new telescope could detect radio signals from another world. For argument's sake, he added, let's just suppose these signals coming from space are equal to the strongest signals we are able to generate on Earth. Well, he said as he produced a piece of paper filled with scratchy, longhand calculations, it turns out that our eighty-five-foot telescope could hear signals as far as ten light-years away. And wouldn't you know it, he continued with almost theatrical casualness, there are quite a few stars very much like our Sun within that distance. Then, apparently no longer able to contain his own mounting enthusiasm, the rest came out in a quick burst: And orbiting around these stars could very well be planets—*populated* planets!

Possibly, said one of the senior scientists seated with him in the booth. But the word was spoken like a challenge and, more telling, he had raised—and this intimidating act would be long remembered—a lone french fry toward Drake as if it were a truncheon. Then in a voice meant to suggest that it was his particularly miserable fate to have to indulge the half-baked schemes of slow, novice astronomers, he asked, "Just *what* is it you're suggesting?" By then all eyes at the table were fixed on Drake.

So he explained. He suggested they begin to search nearby stars for radio signals produced by extraterrestrial civilizations. It would require outfitting the telescope with some new equipment: say, a narrow-band receiver (which wasn't used in those days, and that raised a few skeptical eyebrows) and some amplification devices that had been successfully increasing sensitivity on the Naval Research Laboratory dish—a maser or a parametric amplifier.

Also, he suggested they build their equipment to operate at the 21-centimeter line frequency. It was the precise frequency Morrison and Cocconi had determined as the only "rational" interstellar channel, but Drake did not know this. In fact, as he explained that afternoon to his luncheon companions, he had decided on that specific wavelength because it made the most economic and—no small consideration—political sense. The same narrow-band receivers that would be used in his proposed star search could also be used in another, undeniably significant (and uncontroversial) experiment: to search for the Zeeman effect (the charting of magnetic fields) in the 21-centimeter line of neutral hydrogen. That way, Drake shrewdly suggested to a group of professionals who were bound to be impressed by any magic that would stretch grant money, we would be able to get two experiments for the price of one. Almost daring to share a wink, he added, "Nobody in Washington could snipe that we're wasting funds on farfetched ideas." And, he asserted, "we could build the whole thing for not more than $2,000." On that confident note, his speech done, he turned to look at the man sitting at the end of the table—Lloyd Berkner, then head of the observatory.

Berkner was a pioneer in radio science, a man who had gone to the Antarctic with Admiral Byrd, and he was also something of a scientific gambler. He put down his half-eaten hamburger and stared across the table at young Drake. "Go ahead," he said. "Build it."

It wasn't easy. Drake didn't have to steal to stay within his two-thousand-dollar budget, but he did have to beg, borrow, and improvise. When he had first described the receiver to his colleagues over lunch, it was a simple instrument. It would have only one signal channel and the least complicated of outputs—a chart recorder. And, he had quickly amended, flush with optimism for the inevitable prospects of his still hypothetical experiment, we had better have some sort of tape recorder, nothing fancy, connected to the system to play back the messages we will be receiving from outer space. But Drake found out, as had other scientists throughout history, that between the giddy what-if intimations and the actual flipping of the starting switch fell a shadow of complications.

The design of the receiver became more challenging as Drake refined the description of the sort of signals he was expecting to receive. Stations broadcasting from space (like, say, W-ET) would for maximum efficiency, he now deduced, be concentrated into a band of frequencies no more than one hundred cycles in width. This was an agonizingly narrow range, and Drake ingeniously added some special touches to his "simple" receiver in order to increase its efficiency.

His "souped-up" receiver, therefore, had not one, but two, receiving horns attached at the focus of the telescope's giant dish. One horn was aimed at the target star. The other, pointed in a different direction, collected random cosmic static, the constant stray noises of a bustling universe. At the push of a button, the sounds filling both horns were electronically compared; and, or so the theory went, the difference that would be panned out would be pure extraterrestrial gold—the message from another world.

Also, in a further attempt to eliminate creaking, whistling background noises, Drake rigged up a reference channel device. The receiver, as Drake now had it operating, listened simultaneously on two bandwidths: on a broad bandwidth certain to pick up only cosmic "snow," as well as on a narrow bandwidth where some interesting stations might conceivably be broadcasting. Through the wonder of electronics, the broad-band racket was subtracted from the narrow-band receptions and that would allow the extraterrestrial stations to come in loud and clear. Or so Drake hoped.

But all this inventiveness was nearly abandoned when, six months into building this receiver, there was an unexpected event—Morrison and Cocconi's paper appeared in *Nature.* Drake, at first, was encouraged. Two eminent scientists were not only urging precisely the sort of search he was preparing to conduct, but they were also suggesting that it proceed along the 21-centimeter line. If three people, for whatever reasons, had the same idea, maybe it really was a magic frequency! Then the new director of the observatory, Otto Struve, quickly put things in gloomy perspective.

Struve was furious. A brilliant yet autocratic astrophysicist, as fierce as any commissar from his native Russia, Struve had been one of Drake's most vocal supporters. Yes, he had told Drake, of course there are other worlds. Why, in the Milky Way galaxy alone, he had estimated, there were probably fifty billion solar systems. But when Struve wasn't cheering the project on, he was growling. Hurry up, he had urged Drake. Science is a race. A race for discoveries. A race for credit. A race for research grants. And now with the publication of the article in *Nature* and, more infuriating, all the attention it was receiving, Struve was convinced that the race had already been run and that Drake—and Green Bank—had come in second.

Unless, he decided after a spell of moody, disconsolate brooding, it wasn't too late to announce that his man Drake was also off and running. About a month later, Struve found his chance. He rose to the podium for a long-scheduled lecture at MIT, and as he settled in, he threw out his anticipated script. Dramatically, he read from a new one. My fellow scientists, he announced, the Green Bank Observatory has been preparing, quietly and secretly, for some time now to search for radio communications from other worlds.

This stunning revelation had two consequences. The first was a gift. On a blazingly warm Indian summer afternoon, Drake came down from his office and found that a bear of a man with a long, flowing red beard, Zeus himself if it hadn't been for the tam-o'-shanter angled jauntily on his broad head, had pulled up in an open-topped Morgan. And strapped to the car's front seat was perhaps the most sophisticated electronic device in the world for enriching weak radio signals—a parametric amplifier. The driver was Sam Harris, a self-styled electronic genius who might just have been the real thing; and he had come all the way from Burlington, Massachusetts, bearing this token from his boss, Dana Atchley, president of Microwave Associates and a brother in the search for extraterrestrial life. He had instructions to install the device, fine-tune it so it would magnify the sensitivity of the receiving horns, and then hang around Green Bank until Drake, too, could master all of its arcane mechanics. Which took some doing. But at last, after nearly three patient weeks of early morning tuning sessions, Drake got the hang of things; and Harris, his tam-o'-shanter once more back on his head, jumped into his Morgan and drove off as mysteriously as he had arrived.

The second consequence was, quite possibly, more momentous. There was no turning back. The scientific community, both skeptics and believers, were, thanks to Struve's announcement, constantly harassing Drake. What was happening? they pestered. When would the search begin? they badgered. And Struve himself, excited that Green Bank was back in the thick of things, was not about to allow Drake to lose momentum. Get on with it! he kept on urging.

And then finally, on April 8, 1960, Drake—and the rest of the human race, for that matter—was ready to learn if there were any neighbors, fellow radio buffs, living somewhere in the great expanses of endless space.

He was scrunched in a container not much bigger than a garbage pail. It was perched at the apex of the telescope's focus, ten dizzy stories off the ground. It couldn't have been later than four in the morning, and already he had been up for a groggy hour. And if all this wasn't enough, the container was solid metal so that it caught the howling dawn wind ripping through the valley, and that left Frank Drake, sitting there as good as trapped, shaking from the cold. All in all, he decided, it was a hell of a way to start what might be the most remarkable day in the history of mankind.

But for nearly forty-five minutes, on that first morning of the first day, he sat in the metal can high above the ground patiently twiddling the micrometer adjustments on the parametric amplifier until they were just right, the oscillator tuned in frequency to within one part in a billion. When that was done, he climbed down from the focal point very slowly, one tentative step at a time, careful not to look even once at the hard, flat field below.

Fortunately, there was hot coffee in the control booth, and this eased a million ills. Revitalized, Drake began to adjust the receiver. The equipment was centered on the hydrogen frequency of 1,420.4 megacycles (1,420,400,000 cycles) per second. It would tune into 100 cycles of bandwidth for about a minute, and then automatically move on. The receiver, therefore, would jump across a narrow cosmic dial about as rapidly as if it were in the hands of a petulant teenager twirling his car radio's FM dial as he searched for the one station that he knew had to be playing, at just this moment, his latest fave rave. Which meant receiving a detectable signal was certainly possible; however, probability was a more philosophical question.

Still, as Drake ordered the telescope adjusted, he was optimistic. The giant dish began moving slowly, rotating until it was aimed high above the valley toward the southeast. The telescope found its target, a newly risen Sunlike star, Tau Ceti, and the mechanisms were set that would keep the antenna pointed at the star as it moved across the morning sky. Then Drake began calling out the final checklist. Receiver? Set, was the prompt answer. Tuning motor? Set. Chart recorder? Set. Tape recorder? Set. And with that, Project Ozma began.

Always playful, Drake had decided to name his experiment after the princess of the imaginary land of Oz.

It was an anxious morning. Everyone in the control booth was waiting for history to happen. Each time the pen attached to the chart recorder started to move up, there was a sudden, breathless silence, a moment when everyone just knew *it* was really about to happen.

But then the pen would fall down again, obeying the universal law of Gaussian noise statistics, and the mood—as well as history—would slide away. By noon eight hours had passed, Tau Ceti had set in the West, and nothing out of the ordinary had been received.

Science, like all detective work, is largely a waiting game. At first when the pursuit is still fresh, the clues still infallible, expectations are large. When nothing happens, the edge is lost. Doubts not only intrude, they also for the first corrosive time make perfect sense. Such was the tired, slow pace in the control room on that first day when Drake, shortly after lunch, ordered the telescope to aim at a new target—Epsilon Eridani, another solar-type star. For this search, Drake decided to hook up a loudspeaker to the receiving array. At least, he thought wryly, the constant hissing and whirling of the universe will keep them all awake.

And so the second search of the day began. The chart recorder's needle went up . . . and then it went down. Cosmic static buzzed continually from the loudspeaker. Minds wandered. Eyelids grew heavy.

Then it happened.

The chart recorder started jumping frantically. Bang! Bang! Bang! The needle was knocking rhythmically into its pin. Drake looked at the paper. The pulses were coming in eight times a second. *Every* second. They were uniformly spaced, the product—of intelligent beings. And all of a sudden bursts of noise—not static, but *sounds*—started coming from the loudspeaker. Whooop! Whooop! Whooop! The sounds, too, were being broadcast with machinelike precision: a whistle tooting regularly across the celestial night.

The control room was alive with pandemonium. It was still day one and already they had made contact. Had intelligent life been out there all this time, sending signals for who knew how many years, just waiting for Earth to get smart enough to tune in? Was this the first moment of the future?

Drake, an iron-willed captain, took control. "Let's not rush to any conclusions," he told his bouncing, jubilant troops, though his voice, too, was anything but calm. He ordered that the telescope be moved off the star. If the signal remained, then the source was probably man-made, from Earth. If the signal disappeared, well, then, who knows?

The signal disappeared.

Quickly, his rigid demeanor all camouflage, Drake quietly ordered that the telescope be reaimed at Epsilon Eridani.

This time he heard—nothing. The star was silent.

Was it a fluke? Drake wondered. Or was it really this easy?

There never was a definitive answer. For the next nine days Drake continued to point his telescope at Epsilon Eridani. And for nine days they heard nothing.

On the tenth day the signal returned. This time Drake also held a small antenna out of the observatory window. The signal was being received from another part of the sky by this device, too. And that meant it was man-made radio interference.

Or, at least that was one theory. "We never really knew what it was," said Drake. "We never really knew what we made contact with that first day."

Project Ozma conducted 150 hours of star searches. No other signals were detected. But while Project Ozma never found the proof it was listening for, Drake's mind refused to rest. He was convinced undiscovered worlds were out there. He wanted others to believe it. The search had to continue. That was why he decided to reveal his formula to the members of the Order of the Dolphin.

FOURTEEN.

Even today, when the possessors of the original leaping dolphin pins are treated with reverence in SETI circles, there still remains some debate as to what precisely caused the fraternity to adopt its name. Some of the participants insist the christening came about because of a provocation. Others, a bit shamefacedly, blame it on the champagne. The truth, it seems, lies somewhere between these two extremes.

The circumstances which led to the creation of the Order of the Dolphin began to unfold when eleven wise men, all distinguished names in their fields, were invited by the Space Science Board of the National Academy of Sciences to the remote Green Bank Observatory in November 1961. They had been told to keep the meeting as secret as possible and to a man they readily agreed. No one wanted publicity. The topic for discussion was "Intelligent Extraterrestrial Life."

They were all believers, but on the first day they were served with a challenge. John Lilly, head of the Communication Research Institute in the Virgin Islands, hurled it at his fellow wise men in his opening remarks. Dolphins, he began, "talk." They were an intelligent species that erected no permanent structures, used no tools, possessed no science (at least as man would understand it), and they lacked the ability to use the facial expressions and body language that humans do. Yet, they did "talk." They had a sophisticated culture with an oral tradition. But, he went on with a rumble of exasperation, the human race cannot communicate with them. At least not in any significant fashion.

Then, done teasing, Lilly laid down his challenge. If the human race can't determine a way of communicating with an intelligent species on its own planet, why are the scientists assembled here so certain they will be able to have a dialogue with a more advanced civilization on another world?

It was a provocative thought. And daunting. The mood in the room seemed to deflate. The ambitions of the conference, some of the scientists felt, were all at once too grandiose. But all that was before the champagne corks began popping.

J. P. T. Pearman of the Space Science Board rushed in with the news—and the wine. It had just been announced in Stockholm that one of the men in the room, Dr. Melvin Calvin, had been awarded the Nobel Prize in chemistry. Immediately the toasts began. And as Calvin's accomplishments were heralded, all of them began to realize that they, too, were unique. Their gifts, too, were remarkable.

So while the wine poured freely, as whimsy and confidence brewed in equal measure, they agreed that they would not only take up Lilly's challenge, but would also immortalize it. It was decided that from that moment on these eleven wise men, the founding fathers of American science's commitment to search for other worlds, would be known collectively as the Order of the Dolphin. Calvin, full of the moment, proclaimed that an insignia, an initiation symbol, was needed; he promised to have eleven pins—one for each original Dolphin—made up.

It was in the heady aftermath of this shared resolve that their youngest charter member stepped up to the blackboard and unveiled his formula. It would become known as the "Drake Equation." And it would be their manifesto.

There are certain moments in the life of a scientist when a career hangs in the balance. Not the least of these is the presentation of a new theory to a group of distinguished peers. The opportunity to skid beyond the ingenious to the preposterous is acute, and that sudden descent can mean ruin to a reputation. Frank Drake, pale as an invalid, was aware of all this as his chalk scratched rapidly against the blackboard.

It wasn't simply that the credentials of the audience watching him were intimidating, though that might have been enough. The Dolphins included the already legendary Cocconi and Morrison and his boss Otto Struve, in addition to such world-class minds as astronomer Su-Shu Huang, the imperious Hewlett-Packard Vice President of Research Barney Oliver, and the always inventive and perceptive astronomer Carl Sagan.

His problem, Drake fully realized, was intensified by the audacity of what his formula was attempting to calculate. Still, he continued writing, and when he was done, he turned to face the Dolphins. All the eyes in the room began to scrutinize what Drake had put on the blackboard in neat, large letters:

$$\mathbf{N = R^* \times f_p \times n_e \times f_l \times f_i \times f_c \times L}$$

At last Drake spoke, and with some gravity. He announced that what he was attempting to do in this formula was to suggest a way of organizing knowledge about the universe. The formula was not definitive. It was not inviolable like $E = mc^2$. However, he contended that if one multiplied all the seven factors in his equation together, a scientist would have an educated, sophisticated understanding of what *N* might equal.

And *N*, he revealed, was nothing less than the number of technologically advanced civilizations in the galaxy that were currently capable of communicating with other solar systems.

Then, the room silent with anticipation, he began, moving from left to right, to identify each of the factors in his equation.

He used his piece of chalk as a pointer. R^*, Drake explained, was the rate of star formation when our galaxy, the Milky Way, was born.

This unknown—like all the elements in the equation, the Dolphins realized—was not precisely quantifiable. The best a scientist could do was guess. Yet for anyone mulling the likelihood of other worlds it was clearly a question of vital significance. If one knew the number of "good" stars like our Sun, stars not too old or not too young, that were coming of age each year in our solar system, then one could begin to determine the number of stars that might shine on planets where life might have recently—say, in the last few million years—reached maturity.

But—and this was also, though unarticulated, part of the equation's shrewd design—the question itself was educative. Before any cosmic searcher, Dolphin or novice, could intelligently estimate an answer, it was required to understand, broadly, how stars evolved.

Some fifteen billion years ago, the universe came into existence as a ferocious fireball of pure radiation—the event commonly known as "the Big Bang." It was in these first moments, as the fireball expanded and cooled, that hydrogen and helium atoms were produced. These elements began flying off in slowly rotating clouds; and, in time, the vaporous clouds became heavier, denser, until they collapsed under the weight of their own gravity. Thus, galaxies were formed.

The next evolutionary step was the creation of stars. These first-generation stars used only hydrogen and helium as building blocks. But stars, like mortal men, were born, they evolved, and they died. It was in the death of these first massive stars, when they exploded violently with thermonuclear fission as their hydrogen was exhausted, that heavy elements were created. The colossal force of this explosion "seeded" these elements about the universe.

After about three billion years or so—one-fifth of all time that has existed since day one of the universe—other generations of stars condensed gravitationally from these clouds of new, improved interstellar material. And it was these late-generation stars, stars created out of recycled stardust, that had the elements rich enough to include rocky planets—new worlds—in their orbiting retinues.

(Stardust, in fact, is a part of all humans on this planet. It is the tie that binds our common cosmic ancestry. Every one of the heavier atoms in our bodies, including the oxygen we breathe, the carbon and nitrogen in our tissues, the calcium in our bones, and the iron in our blood—all these came into being through the fusion of lighter atoms at the center of a star or during the explosion of a star.)

Also affecting any estimate of R^* was the nearly stupefying abundance of stars, each one a sun. The prospects were magnificent. Stars occurred throughout space in huge communities called galaxies. The universe contained over a billion galaxies; there were more stars in the heavens than there were grains of sand on all the beaches of the Earth. In the Milky Way alone, the spiral galaxy where Earth is located, there were, roughly, four hundred billion stars.

The sheer number of stars out there, then, would seem to make the calculation Drake was attempting to make impossibly challenging. With all those suns shining in the heavens, how was it possible to narrow down the field to stars that were similar to our Sun, stars that might support a planet with intelligent life?

Fortunately, astronomers have been able to classify stars according to such characteristics as mass, luminosity, and surface temperature.

The scale—indicated by a series of letters: O, B, A, F, G, K, M—runs from hot blue stars to cool red ones. Our Sun is an intermediate-temperature yellow star of the G type.

Further astronomical detective work allowed a star to be classified also according to the different phases of its life history—from its birth as a globule to its death as a fiery, heavy-element-producing supernova. Our Sun is in an intermediary stage, what is called a main sequence. That means it has been around long enough to help life develop, say four billion years. And it will be around for another seven or eight billion years before it explodes—long enough for life to grow up and get smart.

Therefore, a scientist trying to determine the number of stars that at this moment might support intelligent life would restate the question: How many stars are F, G, and K main sequence stars?

The answer—one-quarter of all the stars in the heavens.

Which means there are a lot of potential suns out there.

Drake, quite stoic, his mood as confident as any professor's now that he was rolling, continued to the next factor in his equation. His chalk hit against the blackboard as he pointed to f_p. This was, he explained, the fraction of stars with planets.

For someone searching for other civilizations, estimating this factor was vital. Planets, to use the analogy of one SETI scientist, were "cosmic petri dishes," the controlled environment necessary so that the experiment that will eventually result in life can proceed. Without planets, all available evidence suggested, there cannot be life.

Yet, not a single planet other than the nine in our own solar system has been observed anywhere in the entire Milky Way. The answer to Drake's second question—How many stars have planets?—might very well be only one: the Sun.

Or maybe not. Simply because planets around other stars have not been directly observed with telescopes doesn't necessarily mean they don't exist. Actually, the huge brightness difference between the shining star and the dull planet, as well as the fact that planets give off little heat, made such sighting just about impossible. An analogy: Imagine you're playing center field. Someone hits a high pop fly. You're going to lose that ball in the Sun, too.

Also, more encouraging news, astronomers in recent years have begun to support theories that favor the existence of other planets. This is a change from earlier in the twentieth century when astronomers were holding to the catastrophic theory of planet formation. That mind-set, as its name implied, argued that planets were formed out of the debris from some cosmic disaster such as a sun's exploding or colliding with a third body. If this were true, f_p wouldn't amount to much. Stars are so widely separated that only very rarely in the history of the galaxy would they orbit on a planet-producing collision course. However, in the last thirty years, the nebular theory of planetary formation has come back into favor. According to this theory, planets were the offshoots of the large, rotating interstellar clouds that helped give birth to stars; consequently, companion planets should accompany the formation of every star just as they did our Sun. That would make f_p a whopping number.

And, equally promising, astronomers have been able to uncover circumstantial (yet persuasive) evidence that planets do exist beyond our solar system. There was, for example, the case of "the wobbling dark companion." In 1971 two NASA researchers found that the "wobbling" orbit of Barnard's Star (a run-of-the-mill M star and puny to boot—it had only about 15 percent of the Sun's mass) might be caused by three "dark companions" each as big as Jupiter. More recently, a Harvard astronomer theorized that a large "companion"—about eleven times the size of Jupiter—was pushing and pulling solar-type star HD114762's orbit out of kilter.

So, how many planets are out there? The answer—Drake's f_p—would often be, ultimately, a function of a scientist's personality as much as scientific "fact." An empiricist would insist that all we know for sure is that there are nine planets around our own star. A theorist, however, would make a leap of faith. Most single stars have, like our Sun, their own retinue of nine or more planets; and, therefore, there must be billions of planets—each potentially a life-supporting world—orbiting throughout the vast stretches of the galaxy.

But, Drake explained as he moved on to the next factor in his equation, the problem was not simply finding planets, but also finding planets where life *could* develop. And with that, he pointed to n_e. This required another critical calculation: How many planets orbiting each star had an environment suitable for life?

In attempting over the years to formulate an answer, many SETI cosmochemists, as the scientists who study the chemical origins of the universe are known, have been optimistic—at least in theory. They have pointed out that organic chemistry as we know it on Earth is nothing special. All matter in the solar system has a common origin and the same fundamental elements are prevalent, to varying degrees, throughout the heavens. What is surprising, they decided, is that life arose on Earth of all places. It's a planet that, relative to the Sun, was severely lacking in the volatile elements that made up organic chemistry—hydrogen, carbon, nitrogen. Go anywhere else in our galaxy, and these fundamental building blocks of life were likely to be found in greater abundance. So, some cosmochemists have theorized that since the chemistry necessary for a type of life to evolve certainly existed, then life may very well be widely distributed throughout the universe.

Except, nothing man has experienced firsthand has fitted the theory. Astronauts have returned from the Moon without discovering a microbe of organic matter. The Viking missions that landed on Mars found not a trace of present or past organic chemical evolution. Why? If all the organic elements were available, why wasn't life busy being fruitful and multiplying all across the galaxy?

Cosmochemists begin their explanation with an analogy. Suppose one wants to plant a vegetable garden. The would-be gardener painstakingly clears his plot; gets all the best, guaranteed-to-grow seeds; and, with love and great expectations, plants his precious seeds in neat rows. Yet, since these are busy times, it is not too improbable that even the most devoted gardener might get diverted. But he's not worried. The seeds are guaranteed; nature will do the rest. Come fall, he returns to his garden expecting to harvest his bounty. Only there's nothing.

What did he do wrong? Well, it might have been his choice of a gardening patch—where he chose to plant those absolutely, positively no-fail seeds. Perhaps he chose a spot that was so hot the seeds burned up. Or, maybe his spot was too cold—the seeds froze. Or too wet, or too dry, or too shady, or too whatever.

And similarly, cosmochemists theorized, potential Gardens of Eden were destroyed over time throughout the Milky Way. It was another fact of interstellar life. Chemical evolution is inextricably intertwined with the evolution of a planet's environment. On Venus, Earth's nearest neighbor, for example, the temperatures at the surface are so high that organic compounds such as amino acids, sugars, and nucleic acids could not survive. Without these active ingredients, life didn't stand a chance.

A planet, then, can only "live" if it is orbiting in the proper ecosphere, as SETI scientists call it, of a star. Too close to any sun, the planet will be too hot. Too far, and life on the planet will freeze to a stillborn death. If Earth had rotated in orbit only 5 percent closer to the Sun, a runaway greenhouse effect would have caused it to burn impossibly hot like Venus. Or, if Earth had been positioned just 1 percent farther out in the solar system, there would have been runaway glaciation, turning the planet into another frosty Mars.

So once again, scientists attempting to estimate another of Drake's unknowns would be confronted with divergent choices. Either a planet with the proper ecosphere that would allow the organic chemistry necessary for life to evolve—the n_e of the equation—was a singular exception to all cosmic events, something that happened only once in the universe's lifetime; or, since Earth managed to find a nourishing orbit around its Sun, it was logical and inevitable that any of a billion other planets out there also could.

FIFTEEN.

Drake's presentation suddenly came to a halt. He had been facing the blackboard, when he paused, and turned full-face toward the Dolphins. All at once he looked uneasy. Whatever confidence had been propelling him seemed with this abrupt gesture to collapse. His assurance was lost. He tried to keep his tone spirited, even breezy, but it was difficult. It was now necessary, he confided, to offer, one scientist to another, an admission.

His speech, in substance, was nothing less than a warning. Up to this point, the factors in his equation had been rooted in the admittedly soft turf of guesswork; though that was, he insisted with some force, the legitimate starting ground for much scientific theory. But from this moment on, as he presented the remaining unknowns in the equation, he would be going off into a new, even more speculative territory. He would be leading the Dolphins pell-mell into the professional quicksand of pure conjecture.

But, intrepid, Drake carefully aimed the very tip of his chalk at the fourth element in his equation, f_l; and he held it there for a long, quiet second or two as though it were the brightest, most elucidating ray of science itself. And then he spoke. The question here was—How many of these suitable planets actually develop life?

It was an impossible question to answer. Biologists still cannot fully explain how life came to be on Earth. And if they cannot explain how it happened here, they cannot even begin authoritatively to predict how often nonliving organisms might become living cells on other worlds. Still, if a game Dolphin wanted to make a stab at an estimate, he might review and mull the implications of recent substantiated research supporting the theory of chemical evolution:

It has been determined with some certainty that the organic chemical processes leading to life began on Earth between four and four and a half billion years ago. At that time, Earth's atmosphere was primarily a mixture of hydrogen, nitrogen, carbon dioxide, methane, ammonia, and water vapor. Add some ultraviolet radiation from the Sun. A bit of volcanic activity. And, abracadabra, in the primitive, relatively salt-free oceans, organic compounds began to be produced from simple inorganic molecules.

Two scientists, Stanley Miller and Harold Urey, were even able to re-create a dynamic model of this primitive Earth. First, they mixed in a flask simple molecules like those found in the atmosphere of primordial Earth. Then they zapped this mixture with sufficient electricity to simulate lightning or, in later experiments, ultraviolet rays that were similar to the Sun's primordial glow. And, abracadabra, the simple molecules broke apart and formed, incredibly, amino acids and sugars—the basic building blocks of life.

But it wasn't life. They had manufactured a beakerful of the ingredients used by life, but they still had not made anything that *came* to life.

Similarly, all theories of chemical evolution can only go so far. The sequence of events that caused those organic molecules in the primitive oceans to assemble into that first living cell some three billion years ago are still a mystery. It has remained the one missing link in the chain of creation that ended in man. Until science knows how that happened, until science understands how organic molecules assembled into single cells, f_l would remain a mystery too. For now, all any Dolphin could do was guess.

Minutes later, feeling a bit forlorn, he would recall, Drake had turned to the fifth factor in his equation. The question here, he explained as he pointed to f_i, was— How many life-bearing planets develop *intelligent* life?

Back a few decades ago, anthropologists were by and large supporting vertebrate paleontologist George Gaylord Simpson's view that the evolution of intelligence was so improbable that perhaps it didn't even happen on Earth. Man was a fluke, they argued. These days, though, SETI scientists can draw on recent research into the biological evolution of humans as well as on the cultural development of man and other animals to make a more optimistic case.

Man, they have begun to realize, is not the only intelligent animal on this planet. Chimpanzees, dolphins, even perhaps elephants have complex social organizations and systems of communication. Therefore, if life develops on other worlds—and that's a big "if"—many biologists now predict that intelligence will, given certain circumstances, also likely develop over time.

A firm (and by now even venerated) cornerstonc supporting this belief is naturalists Charles Darwin and Alfred Russel Wallace's theory of natural selection. Over many generations the organism possessing favorable mutations—those that "work" for its survival—will gradually replace those organisms without them. If there's a better way, Darwin and Wallacc suggested, then plants and animals will find it and stick with it.

It is an appropriate theory for this day and age. Pressure, it contends, is not a bad thing at all. Without it, life is an intellcctual dead end. An example: Sitting at the bottom of the ocean are deep-sea echinoderms. They are as happy, presumably, as the proverbial clams. They possess at birth all they need to get by; their lives are lived without either a fear of predators or a shortage of nourishment. However, this unchallenged species will never evolve over time into the sort of creatures that would discover fire or invent tools—or radio telescopes.

Some geneticists even believe that war, at least conventional warfare, is beneficial to the development of intelligence. A creature that must battle more than simply his own environment to survive will, over generations, grow into a more evolved, communicative, and intelligent being.

Environment, the quality of the world in which any creature has to survive, is also a determining factor in the successful development of intelligent life. Earth's diversified climate, with its abundance of inorganic nutrients and relatively stable temperatures, provided, according to some biologists, the perfect world where man's brain could grow. In fact, given a diversified incubator planet like Earth, many exobiologists—as thc SETI scientists who study life-forms as they might occur in an extraterrestrial environment are known—predict that over five hundred million years or so any primitive multicellular metazoan could grow up into an octoploid creature with a human-size brain and the ability to manipulate objects.

And the more diversified the environment, the more ways, biologically and culturally, a creature, according to Wallace, Darwin, and most SETI scientists, will keep on evolving. And the more it evolves, the more intelligent it will become. It worked that way on Earth, so why not on other worlds?

Another case in point: The eye has been "invented" three separate times on Earth—the cephalad eye, the insect eye, and the vertebrate eye. They all have different, totally independent evolutionary histories; yet each of the three organs has evolved to serve essentially the same purpose and all three have basically the same neural networks. So, it would seem that in any world where the optical spectrum band is important, there would be a good chance its inhabitants would develop a light-sensing organ with similar nerve structure. And, the same logic insists, in any world where brains could aid survival, creatures would, nature willing, develop them. Or, at least that is the theory.

So, what is f_i? Are there intelligent beings out there? Everything science knows would seem to support the possibility that if life somehow got started, if the weather was good, and if living on the planet was nasty, brutish, and demanding, it would have been only a matter of time before intelligence evolved.

The next-to-last factor in the equation, Drake stated as he went on cautiously, involved not merely speculations about whether these extraterrestrials were smart, but whether they were smart like us. F_c asked— How many planets with intelligent life develop the ability and desire to communicate with other worlds?

It was a question that was beyond science's authority. Science can point to only one documented case where once a species with some intelligence, manipulative ability, and a complex social organization had come into its own, a technological civilization developed—Earth. On this planet man was able to pass from the Stone Age to the Nuclear Age without any significant biological evolution. But will cultures develop that way on other planets? Will aliens discover the universal laws of physics? Will they be curious enough to apply them?

No one knows.

Perhaps extraterrestrials would logically evolve to look and think and create *somewhat* like we do. But even if they did, it still is a big, crucial "somewhat." Consider a previous example—the eye. Octopi and vertebrates all came up with the same evolutionary device. Somewhat the same, that is. The difference is they cannot see the same things. Octopi can't distinguish mirror images. Or, consider dolphins. They might be wiser, more poetic, more caring than humans will ever be—a higher form of life. But it will always be impossible to build a radio telescope with flippers. And the likelihood of a civilization, no matter how highly evolved, managing to invent radio lessens dramatically if their world is submerged. Or if they communicate telepathically.

Over the years, those scientists who had no faith in the proposed search for extraterrestrial intelligence looked at the problem and recalled Voltaire. He had snickered that if donkeys had gods, their gods would have long ears. So, these twentieth-century cynics taunted that the same might be said about those hoping to detect a signal from space. If astronomers had gods, their gods would have radio telescopes. For the equation now had come down to a matter of faith. Either one believed that f_c was a sizable number, that advanced technological civilizations with the ability and desire to communicate with other worlds were a fact of intergalactic life; or one was certain they are not out there at all.

Drake, aware of the moment, then declared it was time to ponder the final element in the equation. He circled *L*. And he spoke with some emotion. *L* might very well be the most important question for us on Earth— How long will a civilization capable of interstellar communication last?

The chronology, on Earth at least, the Dolphins realized, was not very encouraging. It took man 3.8 billion years to work his way up from his roots as a primeval organism to an intelligent Homo sapiens. Next, after about another seven hundred million years, he finally managed to become smart enough to invent radio—the way to announce his presence to the rest of the universe. Then, just fifty-seven years later—less than a split second in the march of time—he came up with the hydrogen bomb, a way of ending everything.

The moral might very well be that intelligent beings take preposterously long to reach the stage when they can communicate across the heavens, and just at the moment when they attain this level of maturity, they self-destruct.

If this were true, then the history of the universe is a colossal and ironic Greek tragedy. Worlds destroy themselves at the moment when they can reach out to other civilizations. Perhaps in each generation messages were sent out from distant corners of the galaxy, cries of help, or friendship, or cooperation—all from civilizations that no longer exist.

Or, perhaps there will be a happy ending. Perhaps civilizations can learn to live with nuclear science. Since advanced worlds have no other choice, they adapt. Maybe it's not unlike natural selection: mutual cooperation, international governments, planetary peace—all these seemingly impossible, utopian dreams evolved on advanced worlds because such behavior was necessary for survival.

So what is *L*? There is only one fact a scientist, optimist or pessimist, could throw into his speculations. The human race has managed to survive—so far.

Drake was now done, and the Dolphins were seduced. Whatever doubts they had about their ability to estimate specific unknowns in the equation were readily suspended. Drake had offered them a way of thinking about the universe, about other worlds, that was compelling. It held out the promise, a *scientific* promise, that somebody was out there. And so from that afternoon in Green Bank, West Virginia, the Drake Equation became their way of looking at the heavens: their manifesto.

All the Dolphins tried to fill in the blanks. They attacked the equation as though it were a crossword puzzle. Some did their calculations and, like Drake, decided that *N* equals tens of thousands of communicating civilizations spread across the galaxy. Others were more conservative. It was SETI's great game, and they all had to give it a try.

A typical—not too quirkily bold, not too dryly cautious—run through the equation went like this. Let's say, a scientist starting with *R** would begin, that stars form in our galaxy at the rate of one per year; that one-fifth of all the stars have planets; that there are two planets with suitable environments; that life appears on each of these, so the fraction is one; that intelligence develops on each of these life sites, so again the fraction is one; that, say, one-tenth of these civilizations figure out a way to communicate across the galaxy; and, a really big guess, let's say they remain in this state for a thousand years. Therefore, *N*, the number of civilizations at this very moment in the Milky Way capable of communicating with other solar systems, is—and of course there would be the requisite dramatic moment of silence—forty.

And then the arguments would erupt.

Over the years, playing The Game would become part of a scientist's intellectual initiation into the Order of the Dolphin. The Drake Equation would even become part of NASA's creed, underlying the agency's logic in justifying its decision to search for other worlds. And, inevitably, the most hotly debated factor in the formula was always *L*. How long can an advanced technological society exist? Seventy-five years? Five thousand? Forever?

No one, of course, knew. One guess was as "scientific" as another. It was considered by many to be the weakest intellectual link in the equation. Yet as events proceeded over the next decade, if it weren't for *L* and its somber question, NASA would have had no choice but to terminate its involvement with SETI.

SIXTEEN.

In the years and decades that followed the historic Green Bank conference, the eleven original Dolphins continued to theorize and to proselytize; and, in time, a new generation of scientists committed to searching for radio messages from other worlds came of age. SETI, once an outcast, had become respectable. Even NASA was intrigued; in 1974 it funded a preliminary Interstellar Communications Study Group. And then four productive years later, just as NASA was preparing to build a prototype system that would surpass even the expectations of Morrison's, Cocconi's, and Drake's original vision, the government, turning strident and cynical, withdrew its support. This left SETI and its hope-fostered supporters reeling, but it was only a prelude to a larger disaster. The final act—or so it certainly seemed—was played out with considerable gusto on a hot July afternoon in 1981 on the floor of the U.S. Senate.

> SENATOR PROXMIRE: Mr. President, I send an amendment to the desk and ask for its immediate consideration. . . .
> THE LEGISLATIVE CLERK READ AS FOLLOWS:
> Provided: That none of these funds shall be used to support the definition and development of techniques to analyze extraterrestrial radio signals for patterns that may be generated by intelligent sources.
> SENATOR PROXMIRE: Mr. President, 3 years ago, NASA requested $2 million for a program titled "Search for Extraterrestrial Intelligence"—SETI for short.
>
> The idea was that they are going to find intelligence outside the solar system . . . I have always thought if they were going to look for intelligence, they ought to start right here in Washington. It is hard enough to find intelligent life right here. It may be even harder, I might say, than finding it outside our solar system. At any rate, this $2 million would

have funded the initiation of an all-sky, all-frequency search for radio signals from intelligent extraterrestrial life using some of the existing antennas of the Deep Space Network at Goldstone, Calif., and some state-of-the-art hardware that was to be developed specifically for the program. The total cost of the program was to be $15 million over 7 years.

These funds were stricken from the fiscal year 1979 HUD-appropriation bill a few months after I gave NASA a "Golden Fleece" for the proposed project, which I thought should be postponed for a few million light-years.

I have since discovered that the project has been continued at a subsistence level despite our decision to delete these funds 3 years ago. In 1980 NASA spent $500,000 on the project. The 1981 budget was $1 million. NASA plans to spend an additional $1 million in 1982 to continue the definition and development of techniques to analyze extraterrestrial radio signals for patterns that may be generated by intelligent sources.

Mr. President, clearly the Congress intended to stop this research back in 1978 when it terminated the funding for the program. However, NASA has quietly continued the work. . . .

Finally, Mr. President, if we continue to allow NASA to pursue this effort to intercept signals from some hypothetical intelligent civilization, we are sending exactly the wrong signal to the American taxpayer.

We should worry more about improving our ability to communicate with our neighbors on planet Earth and worry a little less about interstellar conversation. In this year of all years we should not fritter away precious Federal dollars on a project that is almost guaranteed to fail. I hope my colleagues will support my amendment to stop this ridiculous waste of the taxpayers' dollars.

THE PRESIDING OFFICER: Who yields time?

SENATOR GARN: Mr. President, on this amendment, the Senator from Wisconsin and I do not disagree. I realize he has a great deal more experience, having been in the Senate a lot longer than I and trying to find intelligence in Washington. . . .

SENATOR HUDDLESTON: Mr. President, on this side, we, too,

> are willing to accept the amendment of the Senator from Wisconsin and commend him for his diligence in ferreting out unnecessary expenditures and seeking to reduce them.

After such back-patting performances, all that remained was for NASA to order the mothballs. Without a dissenting voice being raised, the Senate passed the Proxmire amendment. On October 1, 1981, federal funding support for SETI came to an end. The U.S. government had officially decided not to look for other worlds.

A small, rather muffled parting shot was fired by NASA to mark this solemn occasion. Charles Redmond, of the NASA Office of Space Science, complained to the press, "It means we will have to stop looking at our space shore for a message-in-a-bottle cast out by another civilization. Sadly, if you don't look, you will never find anything." And growing bolder, he offered a wry but still careful response to Senator Proxmire's insistence that "there is not a scintilla of evidence that intelligent life exists" somewhere in space. "As late as 1491," he reminded reporters, "there was not a scintilla of evidence that America existed."

But while NASA officials could walk away from this wreckage with just a despondent shake of the head and then, without even breaking stride, find solace in the burgeoning schedule of space shuttle launches, SETI scientists were disconsolate. The momentous future that the Dolphins had made seem so close was now very far away. Perhaps it was even lost forever. Still, leave it to Frank Drake to find a crumb of humor in the situation. His last laugh made the rounds of the small club of scientists across the nation pondering the possibility of extraterrestrial life; and after each telling, it was invariably applauded with the sort of faintly hysterical irreverence serious men usually reserve only for a wake.

There's a new element, his joke dryly began, to the Drake Equation—*P*. The question here is: How many elected officials does an advanced technological civilization have that are scientifically ignorant? Well, went the punch line, if P = 1 or more there's no need to waste your time estimating the other factors in the equation. All you need is one Proxmire and a civilization will not have the ability to communicate anywhere else in its solar system.

• • •

And yet practically moments after "the fall" (as the Dolphins and their protégés took to referring to their banishment from the rich garden of government funding), there was a sudden, heady hope for redemption. The Soviet Academy of Sciences announced that on December 8, 1981, it would host an international symposium on SETI. This was cause enough for optimism. Certainly Congress wouldn't want the Russians to make contact first; federal dollars, according to this chauvinistic logic, would now assuredly be restored. But there was also the promise of an even greater treasure. A heart-thumping rumor traveled in the wake of the Soviet Academy's terse announcement. There was speculation the conference had been convened to alert the world to a discovery—the Russians had already found something!

It was, considering some of the other wild rumors that gallop through the halls and laboratories of science, not all that outlandish. After all, the Russians, inspired by Morrison and Cocconi's *Nature* article, had for over a decade been using radio telescopes to hunt for extraterrestrial life. And while there were, as might be expected, some differences between the searches at the Gorky Observatory conducted by V. S. Troitsky and Nikolai S. Kardashev (the Russians called their discipline CETI; the day would come when they would not simply search but also *communicate* with aliens) and the pioneering American experiments, there was also a shared, common belief—other worlds were out there. And, certainly giddy fuel, the memory of another announcement was still fresh—the first time the Soviets had claimed they had detected a signal.

On April 12, 1965, the Soviet news agency Tass had reported that Gennady B. Sholomitsky, an astronomer at the Sternberg Astronomical Institute, had observed rhythmic fluctuations in the signals emanating from a location more than two million light-years away in the Andromeda galaxy. This radio source, known to astronomers as CTA-102, was the beacon, heralded Tass, of a "supercivilization." It was proof, the official Soviet news agency proclaimed, "that we are not alone in the universe."

By the next morning, however, the beacon was not shining so brightly. A group of eminent Soviet astronomers, including an embarrassed Sholomitsky, held their own press conference. Yes, they weakly agreed, while it was "possible" that CTA-102 was some sort of an artificial signal, it most certainly was "a little premature" to reach that conclusion so definitively. And before the year was out, the Mount

Palomar Observatory in California had made an announcement of its own. A blue-shining object had been found at the precise location of CTA-102. The radio emissions were not artificial, but were the natural "glow" from a quasar. The Russians grumbled that this still did not explain the apparently deliberate pattern of fluctuations Sholomitsky had monitored, but from that point on there had been no more talk of beacons from "supercivilizations."

Unless, some SETI scientists were suddenly quick to theorize, the Russians were holding back, lining up their facts and figures into neat irrefutable rows, preparing for the Big Announcement. As the American delegation completed the four-hour ferry ride from Helsinki to the Baltic seacoast city of Tallinn, Estonia, the site of the 1981 Soviet Academy conference, such suspicions only grew stronger. These SETI scientists had traveled halfway around the world to arrive in a frigid, glum city just a stone's throw from the Arctic Circle; it didn't take a group of astronomers long to plot that the sun rose each morning at ten-thirty, hovered tentatively over the southern horizon, and then with depressing regularity set each afternoon at two-thirty. These oppressively gray surroundings, they reasoned (or hoped), were only proof that large events were at hand. Leave it to the moody Russian soul to stage high drama against such a bleak backdrop.

But, alas, the SETI scientists had read too much Pushkin. Sometimes a brooding, shadowy city is simply a place where there's insufficient sunshine. Many of the Soviet conference papers were, to quote a surprised and testy Frank Drake, "preposterous." In seminar after seminar, the Russians eagerly threw out bundles of theories that, the American delegation complained with indignant scorn, "violated the basic tenets of true science." Viking probe pictures proved that there are colossal monuments on Mars built by aliens; "blue straggler" stars were dying suns kept alive by an advanced civilization who threw hydrogen "logs" into them to keep the nuclear fires burning; and there was even the suggestion that unusual celestial objects such as star SS433 were the handiwork of creatures from other worlds. The closest thing to a Big Announcement was V. S. Troitsky's (who along with other senior Soviet astronomers seemed rather uneasy as his colleagues paraded their more far-out theories before the Americans) description of the recently launched CETI instrument construction program. An array of a hundred antennas, each a meter in diameter and tuned to the 21-centimeter wavelength, was being erected. He boasted that the first portion of this system, enough to scan the entire sky above the horizon, would go into service within the year.

It was an announcement that clearly demonstrated the Soviets were willing to commit millions of rubles to the development of a dedicated facility that would search space for radio signals; yet, it was not the sort of triumph that was guaranteed to make Congress, to use the jargon of the grant world, come out funding. Still, it would have to do.

It didn't.

"Although I was interested to learn that the Soviet Union is pursuing its own research for extraterrestrial intelligence," wrote Senator Proxmire to Barry Perlman of the Fox Observatory in Dania, Florida, "the mere fact of Soviet involvement does not mean that we would be foolish not to continue our own program. The Soviet Union does not always make wise expenditures any more than the United States does. . . ."

SEVENTEEN.

But the best science is irrepressible. As are the best scientists. The radio telescope searches of the heavens, while not on the scale envisioned when NASA was footing the bills, were not abandoned. The hunt continued.

These were, by necessity, inventive days. In this low-budget period after "the fall," there was a makeshift quality to SETI programs. Just a few years earlier a group of astronomers and physicists, inspired by gaudy visions of NASA funding, had contemplated building a dedicated SETI complex nearly the size of a small city. Cyclops, as this facility was to be known, would be constructed "at the rate of perhaps 100 antennas per year over a 10 to 25 year period" at a total cost of "on the order of 6 to 10 billion dollars." Now, however, the same scientists, sobered and contrite, spoke reverentially of "the Stephens way of doing things."

Robert Stephens, an amateur electrician and astonomer living in the Yukon Territory, had been able to purchase for pennies two fifty-foot antennas that once were part of the Strategic Air Command's northwest Early Warning System. He hammered old tin cans until they were shaped into antenna feeds, rigged up a pulley system out of rusty bicycle chains to help turn his receiver array, somehow put together a working receiver with spare parts from discarded TVs and stereos, and when he was finished he had, at less than what it cost NASA to publish the Cyclops study, a rudimentary SETI system. Yet, however primitive, it worked. Stephens was listening; he could get lucky. Less might be enough.

Likewise at Ohio State University, John Kraus, Robert Dixon, and their construction crew of gung-ho university students built a radio telescope with a flat reflector that was bigger than three football fields for, bargain of bargains, about $250,000. Since 1973 the Ohio State "Big Ear," staffed by volunteer student workers monitoring hand-built or borrowed receiving equipment, had been running nearly full-time searches on SETI targets. And with results—perhaps.

In August 1977, Big Ear had scanned the sky and picked up an intermittent, rhythmic signal. Something was out there—and then it vanished. But it had stayed around long enough to appear on the computer printout, and an excited graduate student circled the telltale pattern and impetuously scribbled in the margins the ultimate accolade a scientist can bestow on any discovery—"Wow!" To this day, the "Wow" signal, as it became known in SETI circles, remains a mystery. Had Big Ear detected the heavenly beeps from a man-made space probe; or had the antenna managed to catch the passing sweep of a beacon from an extraterrestrial civilization?

Such anticlimaxes were, of course, the lifeblood of SETI. But there was no doubt that when the attack came on Big Ear it was launched by humans, specifically the voracious species known as Real Estate Developer. It seemed the university had sold the one site on the planet Earth where an apparatus was searching full-time for other worlds to a developer with his own utopian vision of the common good—he wanted to build a golf course. Mankind, or at least those in Ohio, both he and eventually his lawyers argued, would be better served by the opportunity to play a few holes. Kraus and Dixon, however, refused to surrender. They mobilized students, townspeople, the press (the Men in Green Pants vs. the Little Green Men, was how the tabloids played it), anyone who would listen. The consequences, the professors argued, were nothing short of catastrophic. We would be giving up a chance to learn the secrets of the cosmos, to discover new planets, new civilizations . . . just to play golf. And the professors, to even their own colossal surprise, won. Their search could continue.

As did others. Perhaps the most ingenious do-it-yourself dedicated system was the brainchild of a forty-five-year-old physics professor from Harvard, Paul Horowitz. As a teenager in Summit, New Jersey, Horowitz had built a "talking" robot to win a high-school science fair and with the same boyish dedication and Mr. Wizard skill he had been fabricating elaborate contraptions ever since. Once he wandered into the world of SETI, though, he found his passion.

At a friend's casual suggestion, Horowitz had stopped in at a lecture the great Drake was giving at Harvard. Before the hour was over he was a convert. So intense was this epiphany that Horowitz, always resourceful, managed to retrieve a discarded page of lecture notes where the Master had scribbled out the Drake Equation in his own hand; and next, moving quickly, the treasure lovingly clasped, Horowitz cornered Drake, and cajoled and flattered until Drake had no choice but to autograph it. ("A signed Drake Equation," Horowitz still rejoices. "Someday it'll be one of the most valuable relics of the twentieth century.") After that it was only a matter of time before Paul Horowitz began tinkering on his *own* device to search for signals from other worlds.

He spent months grappling with the design of the hardware, and even longer building the signal processors themselves, each wire-wrapped circuit done by hand. But after about a year and twenty-five thousand dollars he had come up with the most advanced ultranarrow-band signal detector ever built. In one minute it was able to accomplish more magic frequency searches than Drake's Ozma machine could have done in 100,000 years. And, most ingenious of all, it was portable. You could fit the entire four-piece contraption into a suitcase and hook it up to any radio telescope in the world. Horowitz christened his brainchild "Suitcase SETI."

Suitcase SETI was an intricately complex invention, as just a random paragraph from Horowitz's proud monograph *Ultranarrowband SETI* suggested: "The particular hardware implementations we chose used 24-bit rate multipliers to generate a digitally programmed 8.75 kHz (nominal) frequency, which modulated a stable 30-MHz oscillator to generate a 29.99125-MHz (nominal) sweepable local oscillator. This LO mixed the filtered IF input (bandwidth approximately 10kHz) down to quadrature baseband signals, which then passed through 6-pole active low-pass anti-aliasing filters, sample/hold amplifiers, and 8-bit analog digital converters. . . ." And so on and so on.

But in broad strokes—"*very* broad," agreed a slightly uneasy Horowitz—Suitcase SETI, with the flip of a few switches, worked like this. Signals coming in from a radio telescope were searched by a specially designed high-speed analyzer that was capable of sorting through 131,000 narrow channels centered on the magic frequency; simultaneously a Wicat computer displayed these electronic signals as pulses across its screen; and if, among the normal, everyday rows of hunched, electronic hills there was suddenly a towering mountain peak of a pulse—evidence that an artificial signal was being picked up in the midst of all the cosmic static—a deliberately low-key message flashed on the screen: "Notify operator immediately. Possible signal of extraterrestrial origin." And meanwhile, the whole system was also hooked up to a videotape machine the size of a shoe box that was keeping a running archive. "It was," even Horowitz admitted with unself-conscious glee, "a really marvelous toy."

In 1982 Horowitz packed up his "toy" and, like any professor on sabbatical, went traveling. When he reached the thousand-foot antenna in Arecibo, Puerto Rico, he began to unpack. He spent seventy-five telescope hours examining 250 stars. The machine worked perfectly, but he found nothing. Still, no small lesson in those days of sudden institutional poverty, he did demonstrate that SETI—and sophisticated, narrow-band searches to boot—could be conducted on a Radio Shack budget. Let Proxmire award his condemning Golden Fleeces. Let Congress snidely cancel appropriations. The future was still out there, Horowitz's example told a community of dispirited scientists, waiting to be discovered. All it still took was one man with an idea.

And as these searches continued, some explorers went off in new directions—or at least on new wavelengths. In 1971, seventeen years after Morrison and Cocconi had suggested that searches be conducted at wavelengths near the quiet hydrogen line (21 cm), the NASA-funded Cyclops team decided it made better cosmic sense to widen the frequency range: they had discovered an intergalactic meeting place.

Just a bit farther up the frequency band there was another natural emission—that of the hydroxyl radical at 18 cm. So, they reasoned, the wavelength region between the hydrogen line (H) and the hydroxyl radical (OH) provided logical signposts too promising for any advanced technological society to ignore. As their NASA-published report gushed: "Standing like the Om and the Um on either side of a gate, these two emissions of the disassociation products of water beckon all water-based life to search for its kind at the age-old meeting place of all species: the water hole."

And six years later, another NASA-sponsored committee of SETI scientists could barely control their poetic fervor when they, too, urged that the serendipitous "water hole" be considered the primary preferred frequency band for interstellar search: "Romantic? Certainly.

But is not romance itself a quality peculiar to intelligence? Should we not expect advanced beings elsewhere to show such perceptions? . . . There, right in the middle [of the electromagnetic spectrum] stand two sign posts that taken together symbolize the medium in which all life we know began. Is it sensible not to heed such sign posts? To say, in effect: I do not trust your message, it is too good to be true!"

These arguments, one part sentiment to two parts sentimentality, were nevertheless effective. Most of the new generation of independent SETI researchers, the astronomers who took over from the visionary Dolphins, now tuned their receivers to the water hole frequencies.

Yet institutional SETI, high-tech multimillion-dollar NASA SETI, was also to rise from its precipitous fall. It had been brought low by a Congress blissfully ignorant about the prospects and the significance of a search for other civilizations, but this primeval state was soon about to end. The messenger and savior was none other than one of the original members of the Order of the Dolphin. It was Carl Sagan who was able at precisely the right moment to dangle the *L* in the Drake Equation so artfully that the opposition suddenly felt it had no logical choice but to surrender.

The showdown took place in the office of Senator William Proxmire. Sagan went into battle deliberately low-keyed. He knew the enemy, even respected him some, but had few expectations. To attack, to come out fighting, he had strategized, would be instant disaster. So, instead, he circled. The opening talk between the scientist and senator was about nuclear war; this was fertile common ground since both men had publicly agonized over the apparent and increasing dangers. Each approved of what the other had to say. A bond was slowly building. Then Sagan, his timing perfect, struck. Rather offhandedly, he mentioned the senator's opposition to SETI.

"The whole idea is nonsense," Proxmire barked. "If there are other creatures out there, where are they? Why haven't they landed on the White House lawn?"

Sagan kept his self-control. A man of lesser discipline might have thrown up his hands and abruptly skidded into a tirade about ignorant politicians. But not Sagan.

Well, he began helpfully, perhaps the senator doesn't quite understand why it's really very unlikely that Earth should have been discovered. The galaxy, you see, is a vast place. For all we know the nearest civilization is hundreds of light-years away. And then, his audience intrigued, he played his hand. Factor by factor, Sagan went through the Drake Equation. No teacher could have been more patient. No scientist could have been more knowledgeable. And when he came to the final element, *L*, the question about the longevity of technological civilizations, Sagan still did not rant.

But, of course, it was not necessary; the senator, precisely as Sagan had hoped, had followed every subtle hint on his own. Sagan can still recall the crucial moment: Proxmire "had never heard of the question of the connection between the longevity of advanced civilizations and the question of how many of them there are. And before my eyes I could see two neural nets which were in different parts of his brain that had never met, be introduced to each other."

The senator was incredulous. "Do you mean," he asked Sagan, "that if we find some evidence of extraterrestrial intelligence, somebody elsewhere has avoided self-destruction?"

"That's very much what many of us believe," said Sagan.

Well, considered the senator as he began to test the water, if that's the case, if other worlds might be able to teach us how to survive, maybe SETI's worth the investment. "I'd like to think about this problem," he told Sagan at the end of the meeting.

The senator thought about it, and when he was done he concluded that searching for other worlds wasn't such a bad investment of government funds after all. On October 1, 1982, Congress approved a budget line for NASA's program to detect radio signals of extraterrestrial origins.

A new era of SETI was about to begin.

EIGHTEEN.

Early in each new term, Paul Horowitz, the inventor of Suitcase SETI, ran a slide show for his Harvard students. The first slide that flashed on the screen would be a head shot of Albert Einstein. The portrait was, by design, an exaggerated moment. Dr. Einstein's wispy halo of gray hair was flying wildly about and his eyes were bulging with excitement. It was as if Dr. Einstein had begun his day with just a teensy jolt of electricity to jump-start all his complex circuits.

"This," Professor Horowitz would intone with the clear, resonant voice of a man awarding a medal, "is what a physicist looks like."

Then, without delay there would be a telltale click and a second slide would come into focus. The ninety-eight-pound weakling on the screen had a crew cut, horn-rims with Coke-bottle-thick lenses, a plaid shirt buttoned up to the collar, an officious row of pens and pencils lined his breast pocket, and he was wearing his double-knit slacks hitched up to somewhere near his armpits.

"And this," said Professor Horowitz dryly, "is what an engineer looks like."

By this moment in the show, Horowitz's class would be hooting. But these days in SETI circles the same joke was also making the rounds, and it was not a laughing matter. Many of the founding fathers as well as second-generation explorers found its implications a little too relevant. There was a sour feeling that mankind's last great adventure had been transformed. The engineers and machines had taken over.

In time, I learned to appreciate their dismay. After the NASA jet that brought me to the Mojave had swung in low over the looming Granite Mountains and taxied to a stop across from Leach Lake, dry as sand in the winter sun, I was met by Earl Jackson. Dressed entirely in red—red golf hat, red shirt, red pants, red Nikes—Jackson had spent the past twenty-five years living in the desert and working for NASA. He had signed on after a career as a high-school science teacher in his native South and his job, he explained, had evolved over the years. "These days," he said with a sly grin, "I'm a scientist, philosopher, gofer, watchman, and jack-of-all-trades. If it's happening out here, you can bet Earl's got a piece of it."

He escorted me to his blue station wagon "before it gets really hot," though my shirt was already sticky with sweat. Inside the car, the air-conditioning was on high and classical music was booming. Jackson talked nonstop, and drove very fast. As we sped along, he interrupted a monologue affirming his belief in the inevitable success of SETI to point out "tracks made nearly a half-century ago when Patton was out here training his tank corps" and a parched stretch of desert "that's crawling each spring with a sea of migrating tarantulas." After we stopped at a checkpoint where a grim, side-armed military policeman inspected Jackson's NASA identification and wrote down my name on his clipboard, we drove up a small twisting road and, finally, we reached our destination.

It was a white trailer. It looked to me, from the outside at least, like an itinerant's home, although this would be an odd, lonely place to strike camp. The desert sun was like a bellows, warming the dry, heavy air to a 95-degree heat. Only a few bent Joshua trees were rooted on the flat, cracked brown land; it was brutal, relentless country. The setting would have been almost prehistoric, except in the distance, sparkling brilliantly in the full sun, was a huge parabolic reflector antenna—the Deep Space Network's eighty-five-foot-diameter Venus site dish. It was in this small trailer, in a remote corner of California on the 300,000-acre Fort Irwin Military Reservation, that NASA had established its operational base for making contact with other worlds.

Inside, the trailer was as dark and as cool as a cave. There were no people. It was filled with machines. Jackson led the way toward the far corner of the trailer. Lined in a short row were four pieces of electronic equipment. They were ominously nondescript; no telltale wires, no dancing display screens, not even any "start" buttons. Three were waist-high; the last was taller than Jackson. All of them were covered in the sort of bumpy white enamel that appliance manufacturers boast is smudge-free. And, most puzzling of all, Jackson was glowering. It was as if he was staring down an enemy.

"Here they are," he announced, "the prototype SETI instrumentation. These are the machines that are going to allow us to discover new worlds." He, the perfect tour guide, pointed out the Sun 100 Workstation, the power conditioner, the VAX 11/750; and then, as he came to the tall machine, the Multichannel Spectrum Analyzer, he gave it a dismissive slap with his palm. With that liberating gesture his mood slipped further. "Of course," he grumbled, "a lot of people forget about these fancy names. We just call these things the trash compactor, the washer, the dryer, and the refrigerator."

And, after a moment, he added solemnly, "Ask me, maybe we've come too far since the days of Frank Drake."

When NASA dollars started pouring back into SETI in 1983, the decision was made to channel the funds into high-tech R & D. The pioneers had already grappled with the philosophical; the Dolphin scientists had answered the question (at least as well as anyone could), Is anybody out there? Now, NASA administrators decided, it was time to move on to a technical concern: how to design the hardware and software necessary for an effective search.

And as the project focused on building the complex instrumentation to search through the frequencies in the microwave window, new skills were needed. Men and women, working out of the Jet Propulsion Laboratory (JPL) in Pasadena, who had been trained in the narrowly routinized disciplines of electrical engineering, computer programming, and data processing were recruited.

When I, continuing to retrace the itinerary of the UFO Working Group's SETI investigators, made my way south from the Mojave to the JPL, I, like Earl Jackson and Paul Horowitz, also found there was a different tenor to these new troops. They designed high-efficiency antenna feeds, cryogenically cooled preamplifiers, high-resolution spectrum analyzers, and special-purpose signal processors. But all the while they worked without conviction.

Helmut Wilck, head of the Jet Propulsion Laboratory's Digital Projects Group and the man in charge of constructing the spectrum analyzer hardware, was typical. "I have no confidence in SETI, none whatsoever," he said without hesitation. "To me it's just an interesting engineering problem." As it was for Rachel Brady, who worked on the software that will allow the computer to recognize an extraterrestrial signal. She was equally emphatic: "The whole premise behind SETI is pretty fragile. I spend my days working on a challenging signal detection problem. Period. That's what keeps me here."

The labs at JPL were filled with people busy connecting circuits, designing programs, solving engineering problems. On their walls were official notices: to reorder transparent tape one must fill in the code number 7520-00-240-2408 and cross-reference this with the locator number, 807-11-03 EA; the permit necessary to keep a coffee percolator near one's desk was A18-793, and it must be displayed at all times. These engineers proceeded without trepidation, but without excitement. They had inherited the legacy of the Order of the Dolphin, but they have not found the faith.

But there was also, I soon discovered, a new generation of true believers. When I drove north from Pasadena, and headed, as did the Working Group team, to the Ames Research Center at Moffett Field, Mountain View, California, I met Kent Cullers. Forty years old, Cullers is the father of two, and the signal detection team leader of NASA's SETI project. He helped design the algorithms that will allow the SETI computers to locate a message sent across space. He is also blind.

The mistake, he told me, had been made just moments after he was born. It was irreversible. He had been pulled from the womb so suddenly, he was so achingly tiny, his new little arms grasping about like a pair of small parentheses, that the doctors, as was the practice in back-country Oklahoma hospitals in the 1950s, placed him in an oxygen tent. The rushing oxygen filled his lungs and seared his retinas. By the end of his first day alive, Kent Cullers had been blinded.

But now a scientist, he found a power in the darkness. For he believed there was a force in the invisible radio waves that were at that very moment galloping through space. A purpose in these sounds pitched to secret frequencies. A hidden authority shooting through this unseeable universe. And he was confident of his unique destiny. He will see across a deep, dark, and infinite universe. He will see farther than anyone has ever seen before.

"It's my dream," he said with conviction. "And I know it will come true. SETI will work in my lifetime."

. . .

Still, even the faithful would admit, science had moved too quickly. There was too much to know. It was the times in which they lived—and the complexity of its remaining mysteries—that made the Dolphins slightly irrelevant, nearly historical figures in their own lifetimes. A Frank Drake, ingenious and daring, climbing to the peak of his antenna at a chilly, bone-rattling 4:00 A.M. to adjust his handmade equipment, remained a valiant image. But it was an image of the past.

The future, the Dolphins are forced to acknowledge with bittersweet objectivity, will only come about as real-world technology is applied to the world of dreams. In Project Ozma, Drake searched two stars in seven weeks. In *a millionth of a second* NASA's SETI machines, running on their own, can capture more data than Drake did in his entire program.

Drake's Ozma detector had one channel, one ten-thousandths of a megahertz wide. It could, in effect, listen to one cosmic radio station at a time. NASA's Multichannel Spectrum Analyzer, built by the Digital Projects Group at JPL, can tune into a wide portion of the cosmic radio dial and divide it, simultaneously, into ten million narrow channels of one wave per second. The soul of this new, ingenious machine is the SETI Engine, a silicon microchip, only about the size of a nickel, that is more powerful than a giant Cray-I computer; it can perform eighty million arithmetic instructions per second.

Next, after the signals received from space were sorted (or "diced and sliced"; the Multichannel Spectrum Analyzer has been dubbed by SETI wags "the Veg-o-Matic of the Stars"), they would, also in real time, be run through another series of signal detection tests. Specially designed Background Estimation Convolution and Threshold hardware would sift with even greater precision through these ten million "noisy" stations. By now, about four hundred billion data points would have been, with the help of a computer using a threshold detection algorithm formulated by the JPL's software group, reduced to about forty.

Then "Alice" would come into play. Alice, an allusion to the Cheshire cat in *Alice in Wonderland* whose smile (like many radio signals) comes and goes, is pulse detection software. It would guide the computer in its search through these forty "prime" cosmic noises, rapidly cataloging and analyzing the sounds until it found precisely what it was hunting for—a steady, artificial pulse: a message.

This state-of-the-art instrumentation would be, according to NASA's plan, hooked up to existing radio telescopes. The agency, ignoring both the Soviet precedent and the report of the Cyclops study teams, decided it would not build a dedicated SETI facility. There had been, however, some dissension within the ranks of NASA's SETI scientists over where to aim its antennas. It was resolved with a flash of wisdom that struck both sides as Solomonic, especially when compared to the space agency's usual bureaucratic truculence. There would be two complementary searches.

One group, working out of the Ames Research Center, will employ a "target" approach. It plans to zero in on about eight hundred stars in the Milky Way similar to our Sun, using the largest antennas available (such as the thousand-foot dish at the Arecibo Observatory in Puerto Rico). This strategy emphasizes sensitivity; each star will be carefully scrutinized throughout the water hole frequency range. If a signal was being broadcast in this microwave valley, the Ames crew was confident they would hear it.

But, argued the opposing camp of scientists based at the Jet Propulsion Laboratory, such an approach was more guesswork than science; the Ames team, they huffed, was simply wishing on likely stars. So they devised an all-sky survey strategy that, while sacrificing sensitivity, covered the 99 percent of the heavens that the target search ignored. The JPL search, using the 34-meter dishes of the worldwide NASA Deep Space Network, will tune into frequencies that stretch across the entire ground-based microwave window, not merely the narrow water hole.

Full-time NASA sky and target searches, a ten-year, ninety-seven-million-dollar adventure, are expected to begin officially on Columbus Day, 1992. It will be part of the nation's commemoration of the 500th anniversary of Christopher Columbus's discovery of America and, NASA believes, a fitting time for America to discover the future.

In the meantime, the machines in the white trailer sitting in the middle of the Mojave are already listening. Over a loudspeaker, space whistles. It is a shrill, high-pitched burst of noise, a sound that was ,formed fifteen billion years ago, at the moment of creation, and it has traveled across a million billion cubic light-years of dark, vaporous outer space, past superclusters of galaxies, to a green planet orbiting a star on the edge of a spiral galaxy. But it is not the noise the machines are waiting for. They are listening, sorting through the squeaky cosmic hum for a perfect sound, a music not unlike the glorious Bach that first inspired Phillip Morrison—a melody sent across the heavens that will also inspire, energize, and transform life forever. But space only whistles. Yet the music is somewhere out there, loud but unheard, so the machines continue to listen, while men continue to dream.

NINETEEN.

Nearly five months after it had been commissioned, the final draft of *SETI: An Evaluation* was submitted to the UFO Working Group. The NSA investigators had been diligent. The classified report was as thick as a textbook. It included sections titled "Early SETI," "The Logic of the Drake Equation," and "SETI Searches to Date." Its most detailed chapter, however, was "The Prototype SETI Instrumentation." There were turgid explanations of the "36- and 1152-point Discrete Fourier Transforms," complex mathematical equations for charting the "Signal-to-Noise Ratio Loss Due to Quantization," and page after page of "functional block diagrams" and "SETI breadboard data configurations" as convoluted as any map of the New York subway system. The chapter's very heft, so out of proportion to the other sections, did not go unnoticed by the Working Group.

In fact, it was decided that the section was also meant to serve as a "shopping list." The NSA team, signal analysts by profession, had written a section that would, no doubt, be passed on to their mother agency—a glossy catalog of NASA's state-of-the-art signal detection devices. Once this explanation was offered, the excess seemed justifiable. Less easy to explain away, however, was the report's most troubling discovery—the SETI Protocol.

The SETI Protocol was an international agreement that would go into effect the moment contact was made with an alien civilization. It had been drafted by Michael A. G. Michaud, director of the State Department's Office of Advanced Technology, and it was already being discreetly circulated among astronomers throughout the world. Its signatories agreed that any communication with extraterrestrial life was to be "disseminated promptly, openly and widely through scientific channels and public media"; and that an international group of civilian scientists would supervise and regulate all communications with the aliens.

All of which struck the NSA investigators as a "foolish and dangerous" prospect, and they supported that contention with a grim list of the possible consequences of an open and public SETI program. Each was a nightmare the Dolphins, their vision inspired by optimistic science, had never contemplated.

Their first scenario involved the possibility of an alien invasion. Through their efforts to detect signals from space, the NSA authors worried, SETI scientists might very well reveal Earth as a habitable planet. And once aliens knew where we lived, they could embark on an intergalactic crusade to colonize a defenseless human race.

Concern number two was that some advanced race might try to exploit the planet Earth, a concern that, the report noted, was also considered in NASA Publication #CR-11445: "The possibility has been voiced that to a very advanced race we might appear such a primitive life form as to represent delightful pets, interesting experimental animals, or a gourmet delicacy." Though the NASA study also held out the prospect of the human race's revenge: "Differences in biochemistry might equally well make us deadly poisonous."

Next, the NSA investigators raised the possibility of subversion: ". . . an alien culture, under the guise of teaching or helping us, might cause us to build devices that would enable the alien culture to gain control over us. A computer-controlled experiment in bio-chemistry, for example, might be used to create their life form here."

There also was the problem of "cultural shock." Was mankind, the NSA team wondered, really prepared to acknowledge the existence of superior creatures? What effect would this have on ordinary life? Could the human race ever adjust to the reality that it was not the most advanced life-form ever created?

Then, shifting gears, the team's anxieties headed off in another direction. This new concern was designated "The Political Consequences of Contact." If the SETI Protocol were honored, their report noted, the United Nations or an international group of astronomers would represent the planet Earth in all communications with other worlds. And this was clearly "untenable."

For example, the report speculated, what if someone from another world sent a message demanding to be taken to our planet's leader.

That leader, the NSA team insisted, must be the President of the United States. It was not, their report insisted, simply a question of national pride. There were military concerns. Suppose this powerful alien nation could be persuaded to sign a mutual defense treaty with *only* the United States. And there were economic concerns. Suppose the United States could purchase on an exclusive basis, say, antimatter fuel from the alien trade representatives; and in return, neatly restoring the equilibrium to our intergalactic balance of payments, we might negotiate an agreement to sell them all the Pepsi they could drink, all the Big Macs they could swallow.

"In conclusion," the NSA team wrote, "we strongly urge that responsible governmental agencies immediately assume effective control of the SETI program." From the first moment of contact, the report recommended, the government of the United States must "exclusively supervise, monitor, and control" all communication with other planets.

The recommendation was adopted unanimously by the UFO Working Group.

It was toward the end of its discussion of the NSA SETI report that one of the CIA representatives to the UFO Working Group, a longtime employee of the agency's Office of Science and Technology, raised the sole pessimistic note. With all this talk of sorting through the cosmic haystack, he asked with what was recalled as an "antagonistic" edge to his voice, does anyone *really* know what the needle will look like? How will we realize if we have succeeded in receiving a signal from space? Or, even if we manage to isolate a message, how will we go about decoding it?

No one present knew the answers.

It was decided, therefore, to put the NSA team once more on an operational footing. A month or so later they delivered an appendix to *SETI: An Evaluation*. It was a brief report, not more than ten pages, but still it tried to be comprehensive.

It began by attempting logically to scale down the problem. A transmitting civilization *wanted* its signal to be detected; after all, it had initiated the broadcast. The sender, then, would make his signal as visible as possible. It would be nonrandom, pulsations that would stand out dramatically from the noise of space and the rhythms of pulsars.

Similarly, the contents of the signal—its message—would also be encoded in an anticryptographical fashion; that is, in a code that would have been designed not to prevent, but to facilitate, the disclosure of information. An analogy was offered. The cryptologist attempting to decipher an alien communication would be facing much the same problem as the archaeologist pondering an ancient text in a "dead" language. While there would have been no overt attempt to hide the meaning of the message, the key, the intergalactic Rosetta stone, to unlock the cosmic hieroglyphics would still need to be identified.

But, the report reasoned, a cosmic key (or "plaintext" in NSA cryptojargon) was already in the possession of every technological advanced civilization throughout the universe—mathematics. Mathematics, its laws irrefutable and apparent anywhere in the heavens, would, by logic if not by necessity, be the basis of any universal language. It would be the Rosetta stone that would unravel all interplanetary communication.

All of which made perfect sense. Except, as the appendix reported with impressive objectivity, the Drake Message hadn't worked out that way at all.

Frank Drake, the man who had launched Project Ozma and formulated the Drake Equation, had put his keen mind to the problem of designing a concise, easily decipherable "sample" message. The exercise, as he joyfully played it, was part experiment, part science fiction fantasy, and part brain-teasing game. There he was, chief scientist on an imaginary planet, and he was eager to alert the Order of the Dolphin scientists on the planet Earth to his existence. He needed to write a text that would contain pertinent information about his planet, that could be condensed into a series of electronic pulses so it would travel rapidly and economically across the heavens, and that would be readily understood by anyone with a knowledge of mathematics. After Drake had devised this imaginary message, he transcribed the text as a series of ones and zeros (for short and long pulses) and then mailed it to his SETI colleagues.

No one succeeded in deciphering his simple, obvious message.

When Drake finally explained it, they all agreed it was clever. He had sent a run-on series of exactly 551 ones and zeros. This number is the product of 19 and 29. And 19 and 29 are prime numbers; they cannot be evenly divided by any numbers except 1 or themselves. Therefore, 551 equals 19 times 29, and cannot be written any other way.

And once one grasped this, it was only a short, logical leap (or so Drake argued) to another realization—a television picture had been sent. A rectangular "screen" can be built up from the 551 bits by arranging them in 29 rows of 19 characters. Further, once a scientist filled in this screen, putting black squares where there were ones and white squares where there were zeros, a pictograph was formed. The picture showed a crude stick figure of an alien and contained helpful information about its planet's location in the universe as well as its chemistry and biology.

It was very ingenious. A concise, informative message any extraterrestrial would have been proud to send. The sort of message the members of the Order of the Dolphin had always hoped to receive. Except no scientist on Earth was able to decipher it.

Yet the NSA investigators, relentless, managed to give this, too, a shrewd twist. Of course, they grumbled in their appendix, astronomers and physicists could not be expected to break codes. It was a task that required the obscure, finely honed skills of a Fort Meade crypto specialist. Their final paragraph offered a now familiar refrain: Here was another reason to initiate military control of SETI.

Had the government acted on the report's recommendation? Had national security restrictions already been placed on NASA's program to communicate with other worlds? As I began to ask questions about SETI, I could not help but wonder. Indignant NASA officials, however, told me that SETI was "entirely in civilian hands." "If we hear something today," a NASA spokesman assured me, "you'll find out about it on the national news tomorrow."

I recalled that assurance when, later that month, I shared a much too expensive lunch in a pastel-colored hotel dining room just a short walk from the White House with Jonathan Thompson. An ingratiating young Republican with wire-rim glasses and the hearty smile of a game show host, Thompson was coming to the end of a nearly three-year stint as executive secretary to the President's Science Adviser. Part of his job (or his "great opportunity," as he called it) was to report to the Science Adviser and the President on the development of the SETI prototype. It was Thompson who, receiving "red carpet treatment all the way," he recalled with pleasure, repeatedly flew out to the Goldstone complex in the Mojave to meet with the project directors. And it was at his final meeting that he was told by a well-known SETI scientist, "We've been getting some repetitive noises from space. Very curious signals. When Washington finds out, it'll shake them up."

"Forget about Washington." Thompson looked me directly in the eyes and admitted, "That shook me up."

Two weeks later I was in California, sitting in the office of the scientist who had confided in Thompson. I plodded along, doing my best to ask wide-eyed questions about SETI; yet all the while I was waiting for the moment to ask him about his tantalizing remark.

When I finally did, he exploded. "I never said that."

"Well, that's how Jonathan Thompson remembers it."

"He's wrong. He's mistaken. It never happened."

After that, our conversation swiftly degenerated. First the scientist belligerently corrected my grammar. Then, after rumbling caustically for a bit longer, he suddenly decided that there was no reason to discuss SETI with anyone who clearly didn't understand physics. I was asked to leave.

It didn't take me long to get to a phone booth. I reached Thompson in his office in the Old Executive Office Building.

"He wouldn't deny it," Thompson loyally argued.

"Well, he did."

I listened to the hum of the connection while Thompson considered this. "Well, he said it to me," Thompson insisted at last. "I clearly remember him telling me they were receiving repetitive signals."

"That's not what he told me."

"Well, they must have their reasons for not telling you what they tell me."

By this point in my investigation, I was beginning to agree.

Part IV
BOOMTOWN

TWENTY.

It was, I realized all along, the heart of the mystery. I knew the time would come when the members of the Working Group would have to confront the challenge. And yet, I was surprised to find out, no one was eager for the task. UFO sightings confused, troubled, and intimidated the committee.

Part of the problem was the sheer volume of information. Once the Working Group had decided to look for candidates for further investigation, dozens and dozens of sightings began—as if on command—to materialize. From all across America, there were excited reports. Colonel Phillips had collected only the most recent clippings (breathless news articles as well as sober military Flashes) and they filled a folder that was nearly two inches thick. How could the Working Group begin to separate the improbable from the impossible?

There was also, though it was never openly discussed at any of the Working Group's sessions, a bit of philosophy reinforcing their hesitation—they did not want to appear ridiculous. They were all professionals, career intelligence officers or scientists, and, as it was cautiously explained to me, they understood that their reputations were at risk. If they allocated time and money for an investigation into what turned out to be in the end nothing more than a tormented soul's delusion or a blatant hoaxer's plot, their credentials would become equally suspect. No one wanted to be branded a kook, a UFO nut. Yet, the other sharp horn of their common dilemma, it was impossible for the Working Group to walk away from the responsibility.

Of course, the burden fell most heavily on Colonel Phillips. It was his job to find the one sighting that he could present to the Working Group as "highly promising," as the plausible candidate for in-depth analysis. As members of the group waited, the search dragged on, weighing on Colonel Phillips and, they took to noting, perplexing him. He, though reportedly a believer since childhood, seemed for once tentative, unable to make a choice.

For a time, the colonel was convinced he had found his investigative target in Gulf Breeze, Florida. The sightings in this southern town were shared by a variety of witnesses, and there was even a blurred film of what many responsible observers swore was a UFO. But this candidate fell by the wayside after the colonel had some preliminary discussions with friends at Air Force Intelligence. It turned out the Air Force was testing a classified low-flying surveillance plane in the area.

So, with even greater deliberateness, Colonel Phillips once more returned to sorting through the pile of leads; and, in time, he came to another promising candidate—the Hudson Valley sightings. The dozens of UFOs observed streaking through the night sky above a string of northern Westchester communities in New York State shared a very persuasive common denominator—the sightings had been independently affirmed over the past two years by a variety of unimpeachable, upstanding sorts, an eclectic group that included IBM engineers, housewives, and schoolteachers. But then, just when Colonel Phillips was on the verge of recommending that the Working Group dispatch a photographic team to the Brewster, New York, reservoir to join the crowds that regularly gathered for late-night UFO watching, a discouraging report crossed his desk. According to an investigation by a local newspaper, the UFOs were nothing more than convoys of small planes flying with their nighttime running lights on. While this was not an entirely convincing explanation, it gave the colonel pause. Ultimately, he decided the Hudson Valley sightings were not as irrefutable as he had once believed. He realized he would have to look elsewhere. The candidate he delivered to the committee must be unblemished. Too much was at stake.

Finally, early in the summer of 1988, Colonel Phillips convened a session in the Tank to announce that he had at last isolated what he called the "perfect candidate." The witnesses, he explained, were not one or two or even a handful of people—but an entire town.

It was only when Colonel Phillips suggested that the investigative team be recruited from the Domestic Collection Division of the CIA that he, to his obvious surprise, met with an obstacle. The CIA's ranking representative to the Working Group adamantly refused.

"We can't," he announced that morning in the Tank. The National Security Act of 1947, he insisted, prohibited the agency from having any domestic role.

But, he added, even if that weren't the law, the CIA would not be willing to cooperate. The agency's commitment to the Working Group, he explained with a casual dismissiveness that at least one man present thought seemed to jolt Colonel Phillips like an electric current, was a passive one. The CIA would listen, observe, and criticize, but as for sending men out into the field, such participation was simply not part of its agenda. The agency's active concern with UFOs, he stated, had begun *and* ended in 1953 with the Robertson Panel.

Was that really the case? I wondered. The agency's official and public history, I found, concurred with the CIA man's position. Toward the tail end of 1952, the National Security Council had ordered the CIA to, in the words of an internal agency report, "determine if the existence of UFOs would create a danger to the security of the United States." On January 14, 1953, a panel of scientists, CIA officials, and Air Force officers convened at the agency's request for a series of classified meetings (the precursors of the UFO Working Group) at the Pentagon. The chairman of the group's Scientific Advisory Panel was CIA employee and director of the Weapons Systems Evaluation Group in the Office of the Secretary of Defense, Dr. H. P. Robertson. Other panel members included Dr. Luis Alvarez, who fifteen years later would win the Nobel Prize for physics; Dr. Lloyd Berkner, a physicist and one of the directors of the Brookhaven National Laboratories; Dr. Thornton Page, former professor of astronomy at the University of Chicago; and Dr. Samuel Goudsmit, a specialist in atomic structure and statistical problems at Brookhaven.

The Robertson Panel spent twelve hours over three days reviewing about twenty UFO case histories prepared by Air Force Intelligence personnel, two films of alleged flying saucers, and intelligence reports analyzing the Soviet Union's interest in U.S. sightings. The panel's concluding report, classified at the time, was unyielding: ". . . reasonable explanations could be suggested for most sightings . . . total lack of evidence of inimical forces behind the phenomena . . . no evidence of direct threat to national security in the objects sighted. . . ." And its recommendations—whispered advice from one group of insiders to another—helped set the manipulative tone for three decades of government public statements about UFOs: "The continued emphasis on the reporting of these phenomena does . . . result in a threat to the orderly functioning of the protective organs of the body politic" and, therefore, a policy of "debunking" UFO reports should be instigated.

Yet now, thirty-five years later, the CIA was still citing the Robertson Panel Report to the UFO Working Group as the agency's only and final analysis of unidentified flying objects. That gave me some pause: The notion of a twelve-hour roundtable discussion of UFOs as a definitive and scientific study was absurd; the Navy had spent a thousand hours looking at just one of the cases the Robertson Panel had also reviewed, the sighting over Tremonton, Utah, in 1952, and all the Navy's more detailed analysis could come up with was—*object unknown*.

And as I began to investigate further, I discovered that the truth about the CIA's interest in UFOs, though another closely guarded secret, was more complicated. It began at a meeting in a narrow, wood-paneled room in the Brook Club in New York City. The drapes were drawn and the room was appropriately smoky and dim, a breeding ground for mystery and rumors. It was 1948, nearly eight months after Kenneth Arnold's observation of a saucer-shaped craft had created headlines, and the Office of Strategic Services—the World War II intelligence organization that would evolve into the CIA—had convened a special committee to uncover the truth about the flying disks. The next four decades could very well be measured out in secret meetings in dim rooms, in mysteries and rumors, as the OSS and then the CIA continued to investigate and monitor reports of unidentified flying objects.

The history of the CIA's covert preoccupation with UFOs, as detailed by documents obtained through the Freedom of Information Act and in interviews I had with past and present agency officials, was one of obsessive fascination and concern, one that repeatedly pondered the same question—what is out there? And, contrary to the smug certitude in the Robertson Panel Report that the agency has maintained as its official and only position, the actual attitude is less confident. For the past forty years the CIA's classified reports on UFOs have consistently been inconclusive: part deduction, part conjecture, and part, ultimately, frustration.

The paper trail I followed was plentiful—and anxious. Years before the Robertson Panel was convened, a 1949 memo by a Dr. Stone of the CIA's Office of Scientific Investigation (OSI) was full of apprehension about unexplained UFO sightings: "Foreign aircraft development would hardly be tested at such a range from home areas, *even if* fuel could be supplied . . . Guided aircraft at a range of several thousand miles are beyond any known capabilities, including ours. . . ." As was a 1952 internal memorandum from Edward Tauss, then acting chief of the Weapons and Equipment Division of the OSI: ". . . so long as a series of reports remains 'unexplainable' (interplanetary aspects and alien origin not being thoroughly excluded from consideration) caution requires that intelligence continue coverage of the subject . . . It is strongly urged, however, that no indication of CIA interest or concern reach the press or public in view of their probable alarmist tendencies. . . ." Then there were the files filled with "Information Reports" the CIA routinely collected from around the world. The title of each report demonstrated the mounting institutional concern: "Flying Saucers Over Belgian Congo Uranium Mines"; "Glowing Spheres Seen Over Almansa, Spain"; "Disks Appear Over Marrakech"; "Many Witness 'Flying Saucer' Formation in Tarn, France." And so on. Files of questions—and no answers.

The pattern continued. In the years after the Robertson Panel, when the CIA preferred to maintain that its policy was to ignore the phenomenon of UFOs ("the Robertson Report is the summation of the Agency's interest and involvement in this matter," read one public statement), the behind-the-scenes monitoring went on. UFOs were not, it should be noted, a priority concern as they were with Air Force Intelligence or the NSA. But throughout the CIA's Langley, Virginia, headquarters, in carefully compartmented directorates, files were still being quietly maintained. The worldwide sightings reports, for example, were constantly updated. There were dispatches datelined Budapest ("Even if they don't exist, I hope they are on their way to bomb Moscow" was one Hungarian's remark that the CIA field man thought Langley would appreciate); Stigsjoe, Sweden; Bahía Blanca, Argentina; the USSR/Indian border. And there were even files filled with affidavits describing sightings in the United States. One, "Narrative of Socorro, New Mexico Sighting, 24 April 1964," was also discussed in an article in a classified CIA publication, *Studies in Intelligence.* In a paragraph headed "Diagnosis: Unsolved," the article concluded:

"There is no doubt that Lonnie Zamora saw an object which left quite an impression on him. There is also no question about Zamora's reliability. He is a serious [police] officer, a pillar of his church, and a man well versed in recognizing airborne vehicles in his area. He is puzzled by what he saw, and frankly, so are we. This is the best-documented case on record, and still we have been unable, in spite of thorough investigation, to find the vehicle or other stimulus that scared Zamora to the point of panic."

The agency's interest in Americans who had observed UFOs did not stop with Police Sergeant Lonnie Zamora. Throughout the 1970s and 80s, according to statements made by past and present CIA employees, agents from the Domestic Collection Division, using a variety of cover identities, continued to conduct interviews throughout the country. It was a covert project, and, say those familiar with it, it was a loosely run operation. "The interviews were conducted almost on whim, rather than any vigorous agenda," explained a former agency employee. "It was as if the bosses felt it was enough every now and then, when something caught someone's attention, to check out a flying saucer report. But there was certainly no rush to try to solve what was going on." As an April 1976 internal CIA memo noted, "At the present time, there are offices and personnel within the Agency who are monitoring the UFO phenomena, but again, this is not currently on an official basis. . . ." Another 1976 in-house memo reiterated this arrangement: ". . . DCD [Domestic Collection Division] had been receiving UFO related material from many of our S&T [Science and Technology] sources who are presently conducting related research. These scientists include some who have been associated with the Agency for years and whose credentials remove them from the 'nut' variety."

In reality, then, the CIA was neither ignoring decades of sightings, nor had it strictly accepted the easy conclusions of the Robertson Panel. Rather, the agency had by the 1980s reached a somewhat complacent operational level. It acknowledged that no explanation at all for UFO sightings was the only explanation possible.

And in the end, Colonel Phillips won. Whether he went off to Langley and argued his case to a higher and more knowledgeable court, or whether he simply browbeat the agency representative to the Working

Group into acquiescing, remains to this day unknown to most of the group's members. All that is certain is that three weeks later at another meeting in the Tank a scene that had clearly been orchestrated well in advance was played out without a hitch. The ranking CIA representative to the Working Group, trying, some thought, to act as if this was all his conciliatory idea, revealed that the agency could, after all, legally investigate UFO sightings. He cited Executive Order 12333. Issued by President Ronald Reagan in 1981, this amendment to the CIA's charter allowed the CIA for the first time to conduct "special activities" within the United States as long as they did not involve efforts to influence the domestic political process, media, or public opinion. It had been determined, the CIA representative announced as though he were the shrewd arbitrator himself, that the Working Group's purposes did not violate any of the presidential caveats. Colonel Phillips would have his men. To which the colonel responded with a simple and laconic thank-you.

Before the week was over, in July 1988, two agents from the CIA's DCD—there were two of them because sometimes two are less conspicuous than one; and, more shrewd cover, they carried papers identifying themselves as NASA engineers—were sent to evaluate Colonel Phillips's "perfect candidate"—the UFO sightings that occurred in the sky above the tiny town of Elmwood, Wisconsin. And as luck would have it, they arrived just in time for the festivities.

TWENTY-ONE.

There wasn't much to the town. In fact, as I learned when I made my own visit to Elmwood long after the Working Group's investigators had come and gone, the whole place was a small accident of nature. In the late 1850s Abraham Gossett, one of the German sawyers working at the Spring Lake mill up in the Wisconsin north country, decided that he had had enough. He hadn't traveled all the way to America to make someone else rich, so he bundled up what he had managed to save from two years of lumbering (at a princely twelve dollars a week) and, once more, made plans to move on. Along with his pregnant wife and three daughters, Gossett waited until the last snow had melted and then began to head south.

The Eau Galle River was deeper and truly swift in those days, especially after a spring thaw, so he hugged the banks, looking for a safe place to ford. He had been following a winding trail for three days, still hunting for a way across the river, still exploring for his exit to Western opportunity, when he stopped for the night. He never went any farther. More than a century later, there would be a few Elmwood boosters who were quick to give his decision a prideful embellishment: Ol' Abe had realized there was no point in heading anywhere else; he had discovered paradise. A stone's throw from the river, in a small, cloistered valley surrounded by looming, thickly forested hills, he had found what he was searching for.

Another, more truthful way of looking at it, though, was to give the event a more practical twist: He had no choice. His wife had just gone into labor. By the time his fourth daughter was weaned, a log house had been built and the Gossetts had settled in. And by the time those four daughters were married off and living in their own log houses, the town of Elmwood was established. There was a lumber mill down by Cady Creek, a general store at the foot of Norwegian Hill, and even a log church down the road from the Lyss Britton farm. In the census of 1875, the town's population was listed as 730.

Today, Elmwood has grown—somewhat. According to the most recent count, there are 991 people living in the township. I found it about the length of a good jog. Coming into the town on route 72, the road begins to dip as you enter a valley and the first thing you see, standing in a small clearing in the midst of some thick woods, is the WELCOME TO ELMWOOD sign. The greeting is superimposed over an outline map of Wisconsin and there has been some talk of putting a star in the northeast corner, near the Minnesota border, to signify roughly the town's location, but no one has gotten around to it yet. Then, the road begins to elbow around the Eau Galle, more a tranquil, blue-green stream these days, and on the site, more or less, of the old Gossett homestead is the Elmwood Area School. The school is a red-brick box, home to grades one through twelve, maybe one hundred kids in all. Past the playing field, and beyond the majestic, stark white concrete grain tower of Elmwood Feed & Supply, the road straightens out as it heads over White Bridge. The Eau Galle here is deeper, and in the heat of the summer a few bold kids will, after checking to see that Police Chief Gene Helmer is nowhere in the vicinity, take a swan dive off the railing. Once you're across the bridge, and if you continue past Thompson's Gas (one of the three stations in town), and keep on heading toward the stop sign (which nobody seems to pay much attention to anyway), you'll be at the head of Main Street.

Downtown, the locals call it. There are a couple of food stores; in the fall, the deer hunters keep the butcher at Sailer's Grocery pretty busy. Then there's the First State Bank, a quiet place with a couple of tellers behind a screen and a lone guard. For reasons no one is really sure of, the cement sidewalk begins to rise right outside the bank, but by the time you're farther down Main Street and standing in front of Village Hall it's level again. Village Hall, a boxy 1930s red-brick structure with a pair of carriage lamps flanking the glass doors for a bit of decoration, is the biggest building in town. Besides providing offices for the mayor (actually, he's officially the village president, but everyone calls him "Mayor") and the town clerk, it's also home to the library and a second-floor auditorium where the town meetings take place.

(A few years back, someone noticed that not many people were showing up for these town meetings and it was suggested that at least some of the auditorium space might be put to better use. So, after a municipal referendum, it was agreed to convert part of the auditorium into a two-lane bowling alley. It's been a popular decision. Most of the stores and feed cooperatives in town sponsor teams, and the league competition gives people something else to talk about besides the weather during the long winter.)

The rest of Main Street is mostly bars—there are a lot of bars; and people in these parts, perhaps a bit embarrassed by this abundance, prefer to call them taverns. The two most popular places downtown seem to be the Shack, where the kids hang out, and Diane's Sandbar. By three the Shack smells strongly of hot pizza and hamburger baskets, and John Cougar Mellencamp ("Small Town") is wailing from the jukebox.

Over at the Sandbar, things don't pick up until later. About a year ago, Diane put a waist-high front of siding shaped and colored to resemble fieldstone over the red brick, but nobody seems to have noticed. The crowd here, mostly men wearing Elmwood Feed caps and displaying huge, sagging bellies they have carried for so long that they even have names ("Me and Gus'll be going now," one of the farmers will say at the end of the evening as he pats his paunch and lifts himself—and bouncing, bountiful Gus—up from the barstool), come for the talk, not the atmosphere. And to drink. The "usual" is a draft and a shot, though if you're hungry Diane will put a plastic-wrapped bright pink wiener bundled up in a pale bun into the microwave, and zap it for you.

Other than the Black Bear Inn, where Jim Baier's shrimp basket with potato wedges is a tradition, that's really all there is downtown. Except, of course, for the white-steepled United Methodist Church, which while technically "downtown," no one likes to think of as being part of what the Chamber of Commerce calls "the business district." The other churches, in fact, are by design not on Main; Sacred Heart, which just celebrated its seventy-fifth anniversary, is over on County Road P, and St. Peter's Lutheran, where the descendants of many of the town's old-time Scandinavian families pray, is on Shaw.

And once you go past St. Peter's you're starting to head out of the valley, back up into the hills, and out of Elmwood.

People like living in Elmwood. By the early 1900s, the hills had been depleted of their hardwood elms and oaks, so the mill closed down and it became necessary to find another way to make a living. A lot of families began clearing their land into pasture and buying livestock. Dairy farming is the predominant business these days. Over the years, the farms have been passed on from father to son. That's part of the reason residents feel bound to Elmwood; it's their heritage and it will be their legacy. It also helps that it's a town comfortable with tradition. Homes have been known by the same family name for generations. Week in and week out, housewives cook the same meals on the same nights. You can walk down the village streets and the lawns are always neatly cut and laundry is hanging on a backyard line, flapping in the breeze. And, no matter what the time of year, the first topic of conversation is usually the weather.

Life is full and rewarding. In the winter the bowling league keeps you busy; come spring there's trout fishing in Cady Creek; in the summer there's the softball league and that, as anyone will tell you, is a friendly game, only just be sure to wear your spikes; and in the fall, it's hunting season. Sure, there are problems. The cows might go dry. Or you might wake up in January and turn on the radio only to hear a dreaded Traveler's Advisory—Snow Alert; and when it snows in northern Wisconsin, it *snows*. But, as long as you've got your health, the days are busy with hard work, family, and lifelong friends.

It's a town where generations of families have been christened, married, and buried in the same church. The only tie to the outside world (unless you count the mall up in Menomonie, eighteen miles to the north) is television; and what with cable, CNN, and blaring MTV, some people are thinking the rest of the world is getting maybe a little *too* close to their valley of farms and picket fences. Still, life in Elmwood would seem at first glance to be as regular as the seasons, a rare corner of America that has had the luck to ease itself through the intrusive tumult of the twentieth century, a town where day in and day out life goes on without surprises.

Only this is not true. Something is peculiar. Odd. Even extraordinary about life in Elmwood.

There are clues, if you look and listen closely. In the field across from the Elmwood Area School, some of the younger children are playing tag. Julie Walker is "it," only the way these kids are playing the game, red-headed Julie is captain of an alien spaceship, and if she tags you,

you're captured—you've got to fly off to a distant galaxy with Julie and her crew. While over in the school gym, the pom-pom girls are practicing. They're wearing short maroon skirts and white sweaters emblazoned with maroon Es and they're waving their pom-poms as they chant:

We'll zap you with our ray guns;
We're part of the Force.
One blast of our cosmic power
And you'll go sour!
Yea Elmwood!

Or go downtown. In the library on the main floor of the Village Hall, Joanne Baier, the librarian, was getting so many questions, so many requests for the same sort of book, that she decided to make a special section. It's up front, across from the *Encyclopædia Britannica*, and the volumes in this special section fill three shelves. Above this bookcase, in big block letters, is a sign—UFO BOOKS. Or go upstairs, into the town clerk's office. It's all locked up in Dolores Wilson's filing cabinet—the report, issued by the Pierce County Economic Development Corporation board, for this year's annual UFO Days weekend. They're expecting maybe five thousand people this year, Police Chief Helmer has been told.

And down the street at Diane's Sandbar, Hobbs Wilson, still a newcomer to some people since his family didn't move into town until 1914, is going on again about the time he saw two of them the same night by the Spring Valley dam. "Couldn't have been more than two hundred feet away from me," he's saying to anyone who wants to listen. While at the Shack, some of the teenagers are making plans for the evening. Greg Radtke got the keys to his dad's Chevy, so he's going to take Carole and maybe another couple and then they'll meet up with Bobby Thompson, who's got his own truck, and who knows who else, and finally, in a convoy, they'll head on over to County Road P. Then they'll go cruising. They'll just drive up and down the dark, back country roads for hours. Cruising—and waiting. Watching the night sky. Watching for *them* to return.

Hardly anybody in Elmwood doubts they will. Because just about everybody has seen them, or at least has a brother or sister or father who has. Oh, maybe they won't come tonight. But tomorrow or the next night or the night after that—they will. Because year after year, sometimes once a month, sometimes less frequently, this tiny town in an isolated valley in the north woods of Wisconsin is, the locals are convinced, visited by alien spacecraft. No one knows why; and by now, it's gone on for so long that hardly anyone in Elmwood thinks it's really even a big deal.

TWENTY-TWO.

The house was dark. People were home—somehow he knew that—but they were hiding, pretending they were out. So the car (or was it a bus? it seemed bigger than a car) had no choice but to keep on going. It continued at a steady speed down the long, empty road. At last, there was another house. It was a wonderful house. There was a picket fence, and a garden, and blue shutters. A pie, or maybe a cake, was baking in the kitchen and the warm fresh smell was tempting. No, the allure was stronger than the simple tug of hunger. It was intense. Enticing. But this house, too, was dark. Again, everyone was hiding. Why? Why didn't they invite the people in the vehicle inside? Why didn't they welcome them? Yet the house remained dark. Nothing moved inside. The vehicle continued down the blacktop road. It wasn't right.

Tom Weber woke from this dream in the grip of panic. All at once he felt defenseless. He was fifty years old, a husband and a father, a good provider, and for the first time in his very ordinary life things were out of control. He wasn't scared, but something, he desperately understood, was wrong. The digital clock read after 3:00 A.M.; his wife Colleen was cuddled on the other side of the bed, peaceful and oblivious. He wanted to hug her tight, to go back to sleep, but he realized the dream would only return. His escape would be short-lived. He didn't have the will to endure it one more time: the heavy sweats, the unfocused anger swelling from somewhere deep in the very center of his being, the infuriating confusion. What was happening to him?

The dream had started perhaps six months ago. It was always the same: the dark houses, the unwelcome guests, the sharp and corrosive feeling that this inhospitable behavior was an irrevocable mistake, the seeds of something tragic. At first the dream had come upon him casually, maybe once every week or so. But now it was relentless, an alarm sounding in the depths of each night. The time had come, he acknowledged to himself, to understand what he was being told—and to deal with it.

Not that from the start he didn't have his suspicions. But he preferred, as was his methodical way, to do things slowly, to follow a logic of sorts. So, sitting upright in his bed, his wife soundly asleep by his side, Tom Weber began an inventory of his life. In the cozy, homey quiet of the early morning—a false calm, he knew—he forced himself to work his way down a short checklist.

Work? Well, things were just starting to come together. A few years back he had started up a company to spray paint computer parts and even though headquarters was a warehouse in Chippewa Falls, Wisconsin, it was beginning to attract the attention of some of the big computer manufacturers. Jobs were pouring in. Now, if only he could manage to keep his lungs clear of those damn solvents from the spray paints . . . And then it struck him. Perhaps some sort of toxic gas from all the spray painting he had been doing had already been absorbed by his system. Perhaps the noxious by-product was eating away at him; maybe that was what was driving him toward this agonized confusion. It was a chilling thought. But no, he finally decided, his symptoms would have been different. Delusions, yes; a variety of wild thoughts, possibly. But not the same small, precise dream night after night.

So, he moved on to consider the rest of his life. He was a happy man, very much in love, and very much loved. His three children were a blessing, especially the three-year-old. Who would have imagined, he asked himself with delight, that he would father another child as he got within worrying range of his fiftieth birthday? And, another unanticipated bit of good fortune, Colleen was willing to put up with a lot. She was twenty years his junior, still sparkling with youth, and he, with his gray Old Testament beard and big balding dome of a head, looked like her father. At least that was what many people thought when they were first introduced to the couple. And his kids? Most people made the same mistake there, too. Nice grandchildren you have, Mr. Weber. But did Colleen complain? Get embarrassed? Not his Colleen. She was with him all the way, "supportive," he was fond of saying. In fact, there was only one provocation that would summon a streak of her Dutch temper to rush to the surface. "You and your damned science," Colleen, shrill and exasperated, would shout after he launched into another of his lectures. The whole thing made her nervous; she didn't like to think about it. But he did. For years he had been living with these ideas, continually collecting facts, doggedly building theories. He knew he was right. And, as he replayed a familiar scene in his mind—Colleen telling him she'd had enough, she didn't want to hear about it anymore, and he, calm but adamant, reluctant to give in—Tom Weber realized at that moment what he had suspected from the start—the dream was about *them*.

Over the years, Tom Weber had become convinced the human race was not alone in the universe. Science and logic, he insisted with the rigidity of a man who was proud of his self-taught grasp of both disciplines, provided irrefutable proof. Don't tell me, he would begin as he launched into a carefully formulated exercise in probability theory, that it's unlikely there are other worlds. Why, what are the odds that any of us should exist? That our parents should meet? That our grandparents should somehow have been brought together? And that challenge thrown, he would commence tracing a genealogical tree whose branches stretched back into the Middle Ages, an exercise that splendidly demonstrated that if the rules of strict probability truly governed this universe, then none of us would have been born. The odds that all the precise sets of circumstances that would have had to happen to make any one of us the heir to the specific family history we inherited are greater than the number of stars in the universe. "A thousand years ago, who would have predicted me or you?" Tom Weber, triumphant, would ask. "Yet," he would conclude with a last-laugh smile, "we all exist. And, likewise, so do other worlds."

Then, rolling on with perfect confidence, he would launch into his theories about unidentified flying objects and the creatures who piloted these craft. To him, all his deductions were simply common sense. A Super-Intelligent Society was presently monitoring Earth. Since they were so technologically sophisticated, these creatures lived a subterranean existence on their own planet; living on a surface exposed to myriad cosmic dangers, these wise creatures would have undoubtedly learned, was too precarious a situation. Also, because they were so advanced, they employed a mechanics quite different from our own one-dimensional aerodynamics; these creatures—and here Weber could be a bit long-winded, offering his unique, didactic revisionism as the alternative to the axioms in the dozens of physics texts he had so laboriously studied—were able to manipulate the effects of gravity on their craft through electromagnetism. Above all, he would also take care to emphasize, it was impossible that these extraterrestrial visitors were anything but benevolent; for, he would state with unwavering confidence, "the level of a society's technology will always be proportional to those moral standards governing its use."

He had worked out these ideas over more than a decade of deep thought. Yet, as he sat upright in his bed, he could not understand why these carefully rendered articles of faith were suddenly turning on him, suddenly tormenting him. What had happened to turn his lovingly formulated philosophy into this nightly avenging dream?

And then he knew—Elmwood.

It must have been over a year ago when he first started reading in the local papers about the frequent sightings in that tiny village almost forty-five miles from his home. Though a believer, he had never seen a UFO. Perhaps this was his chance. So, when things slowed down at work, he drove to Elmwood. He sat parked in his car on a country road watching and waiting. He saw nothing. But he did make it his business to speak to many of the people who had witnessed these craft. Everyone had a story to tell, and he, intrigued, listened. He returned home disappointed, but still convinced. And that was that.

Or was it? Because now it all fell into place. His trip to Elmwood had been six months ago. *Just when the dream had started!* He rose from his bed as if summoned. The first weak rays of sunlight were sneaking into his bedroom. The muffled sounds of his children's first stirrings were floating through the house. As he stood by his bed, a man erect as if at attention, unable to move, unable to speak, the meaning of his troubling dream was at once apparent. And the odd, unconnected ideas that had for so long been rumbling through his daily life had in an instant acquired a monumental purpose.

From that first moment, the dream, now decoded, unfolded before Tom Weber with a remarkable clarity. The dark houses, of course, were the homes of the citizens of Elmwood. They were the ones who were hiding, refusing to welcome "the vehicles" filled with the superintelligent beings. And he also understood what was making his pain so acute, why his nightmare was such an agonizing one: This inhospitable behavior would have terrible consequences. It wasn't simply that for the throwing open of a door, the turning on of a light, the opportunity to take part in the greatest event in the history of mankind was being ignored. No, there was more. These benevolent visitors were bringing a precious gift—the wisdom they had acquired that had allowed them to survive despite their knowledge of nuclear technology. Without this wisdom, mankind would, he was certain, self-destruct. Nothing less than the future of the planet was at stake; and Earth, or at least Elmwood, was uninterested.

Who could blame Tom Weber if he couldn't sleep?

But, undaunted, he came up with a plan. The details took some long and deliberate pondering, but the broad strokes came to him, he would claim, in the quiet of that first morning when his dream was suddenly translated into a coherent message. Still, he waited awhile before sharing his idea. It was a vast enterprise he envisioned, and he wanted to get it all laid out in his mind before proceeding.

When Tom Weber was more confident, on a blustery February day in 1988, he decided to act. But there are nightmares that do not go away with daylight, and as he drove east on snow-slicked I-94, his mind, too, was swirling like the wet, falling snow. He was once more full of familiar anxieties and terrors, and only as he got closer to Elmwood did it even occur to him that it might have been wiser to have called the mayor and made an appointment. After all, not every politician might have the time to see a visitor—even if the unexpected guest was bringing an idea that would change the world.

TWENTY-THREE.

There are some politicians who are most comfortable with superlatives; the language of excess seems to suit their ambitions. Others are by nature more laconic; or then again, perhaps they simply have something to hide. But Larry Feiler, the mayor of Elmwood, was most likely to lapse into the sort of discursive understatement that left you wondering if he was a tad slow—or whether, just maybe, straight-faced Larry was putting you on. A case in point. Larry, do you happen to know how Diane's Sandbar on Main got its name? Well, he replied after a moment of pensive meditation, could be because the soil in town is so sandy. A beat and then the slightest trace of a grin. But maybe this has something to do with it. Diane's last name, now that I think of it, is Sand. And so when Larry took to describing his first encounter with Tom Weber it, too, was presented with a similar subtle charm.

It was a typical day, filled with typical mayoral problems, Larry remembered. First, there was the uproar over the bowling league. A score from one of last night's games was still being hotly contested and this had provoked a lot of nasty name-calling. Who knew where a lingering sore like that could lead once the bars started packing the farmers in; perhaps Chief Helmer should be alerted? Then there was the naked snowman—or, more precisely, snowgirl—situation. It seems last night someone had artfully sculpted a few distinctive curves and bumps into the bland snowperson across from the bank and in the light of day it clearly was one hell of a snowgirl. Dealing with the offending artwork was no problem; it had been summarily whacked in half with a shovel. But some of the shopkeepers were still demanding that the degenerate sculptor or sculptors be tracked down and prosecuted to the full extent of whatever law was vaguely applicable. And just as the mayor was mulling that one over, in walked a man, unannounced, who said he wanted to build a fifty-million-dollar landing field for UFOs near the town. All in all, said Larry Feiler, a typical day for the mayor of Elmwood.

Even without the benefit of the mayor's ironic narrative gloss, it must have been a very peculiar encounter. By the time he had climbed the two flights in Village Hall and bullied a complaisant secretary into believing that he absolutely, positively had to see Mayor Feiler, Tom Weber was once again brimming with confidence. He was a large man, bulky but solid, and with his gray-white beard and prominent head, he was a presence. Oddly, despite his size, there was a softness to his voice, not to mention a bit of a twang, too. And he was a very verbal man, proudly articulate, and his words flowed in quick, easy torrents. Most striking of all, though, were Tom Weber's eyes. They were a brighter-than-real green. Piercing, people said. As he took a seat in Mayor Feiler's office, those green eyes began sizing up the man on the other side of the desk.

What he saw did not, at first glance at least, appear to be a kindred spirit. Though a life-long resident of Elmwood, Feiler, forty-six, part-time mayor and full-time insurance agent, had a city slicker's polish to him. There was nothing hayseed in his carefully put-together getup. He was partial to blue blazers with the sort of shiny brass buttons that might please an admiral, silky rep ties, and double-knit trousers that flared out just the slightest bit above his loafers. He even wore tinted glasses with gold frames. And while many men in Elmwood had a haircut that looked as if the wife, with the help of a mixing bowl, had taken it upon herself to even out the raggedy strands on a slow Sunday night, Mayor Feiler's bright silver locks must have been blow-dried, razor-cut, and then combed with an artist's precision to get them to lie in such careful, handsome neatness.

Still, after Weber introduced himself, the mayor, despite the urgency of the day's unsolved and lingering problems, could not have been more agreeable. He simply sat back and told the unexpected visitor to "fire away." "That's my job—to listen," he confided with good-natured equanimity to Weber.

And listen he did. If Weber had thought about telling a convoluted story beginning with the perplexing mystery of his recurring dream, that strategy, after a sizing-up of the mayor, was abandoned. Instead, he started in with the ending. I know why the flying saucers are coming to Elmwood, he announced. They want to land.

Then, for the next thirty minutes or so, Weber, mixing his home-grown science with his missionary conviction, outlined his plan for building a two-square-mile UFO landing field just outside "down-town" Elmwood. When he was finished, it was the mayor's turn. There was a lot he wanted to know. Feiler's questions, by his own description, were pointed. And while no tape of their conversation was made, of course, a similar interrogation between Weber and another interlocutor, Marcia Nelesen of the Janesville *Sunday Gazette,* gave, both the mayor and Weber agreed, a good feel of the substance of the grilling that took place that afternoon in Village Hall.

Q. Why do aliens need a landing site?
A. The only collective message that mankind has ever sent them was to try to shoot them down.

There is an excellent possibility that, if roles were reversed, we'd respond to our message. We have nothing to lose by making the attempt and possibly everything to gain.
Q. What message?
A. It is actually a visual message of a human figure, one-fourth mile in length, that will be well lighted. The alien figure [shaking hands with his human brother] is based on an international consensus of what these people look like.

It is plain and straightforward—"Members of your society and our society will meet here." It will be displayed on the landing site in lights.
Q. What will the landing site look like?
A. It will be approximately two square miles and will be used for transmitting the visual message. Some of the space will have scientific laboratories. We intend to invite scientists from all over the world when it occurs. We're going to need labs and housing. There will be a light border around the entire complex.
Q. Why are you convinced the aliens would be friendly?
A. If they weren't friendly, and if their goal was to enslave man or take the planet away, we think they'd have done so already. If they aren't friendly there isn't a damn thing we could do about it anyway. . . .

> We have chased them and shot at them and they have not taken retaliatory action. I think they're waiting for this invitation as evidence that we've learned something and that our attitudes have improved.

And when he was done asking his questions, Mayor Feiler took a moment to consider what he had heard—and, perhaps more important, what it would mean to the town. "Sounds like a good idea to me," he said finally, as he extended his hand to Tom Weber.

Before Weber drove out of Elmwood that afternoon, Mayor Feiler had pledged his support. He had promised to do his best to help Weber obtain the land and, further encouragement, he was going to present the plan for the construction of the UFO Landing Field at the next session of the Village board. "I'll see if I can get them to help," he promised.

But what was perhaps even more remarkable than this immediate alliance between the long-tormented visionary from Chippewa Falls and the practical insurance agent/mayor from the tiny village of Elmwood was that neither man *ever* had any doubts about the reality of UFOs. It was the one question that was never asked, the one possibility that was never raised. Both men took it for granted that alien craft were flying over Elmwood. For Tom Weber, it was a matter of faith. For Mayor Feiler, it was simply a matter of trusting his neighbors.

TWENTY-FOUR.

Seeing is not the only path toward believing. In a small town like Elmwood, the fact that your neighbor had seen something was often good enough. The mayor was a case in point. "I have never had a sighting myself," Larry Feiler would breezily admit. "But," he would quickly insist with undiminished authority, "as far as the people themselves who saw them—I knew Carol Forster and George Wheeler. There's no question about the credibility of their stories. Something is out there."

Everyone in Elmwood, in fact, knew George and Carol; and after what had happened to them, it wasn't just the mayor who, however secondhand the evidence, was convinced something was out there.

George Wheeler was Doris's husband. At least that was the way most people in Elmwood still referred to him. It wasn't meant as anything dismissive; it was just a way of establishing his pedigree. After all, George had only been living in town for seventeen years. Doris, though, was a native; her family had been living in Elmwood *forever*.

George had come to Elmwood after ten years as a state trooper in New York, and because of his training and experience—in addition to the equally important imprimatur of his relationship with Doris—he was quickly offered a job in the town's police department. His hiring doubled the size of the force.

That was back in the 1960s, and for the next decade or so George Wheeler earned the reputation for being what people in Elmwood call "a good guy." He was conscientious about his work, a family man, and, no small matter out in the country, people also grew to respect his toughness. He had the heft as well as the gnarled, menacing look of a once nasty linebacker now turned no-nonsense coach in middle age, and that came in handy whenever a couple of the farm boys got out of line after a few too many beers. George, without a moment's hesitation, would jump into the middle of any shoving match down at the Sandbar, say, and somehow manage to wrestle the guys apart. "Now don't you two boys think you should call it a night?" he would suggest, eyes as hard as bullets, as he stared one, then the other down. It wouldn't be long after that before the two battlers, looking a bit sheepish, would mope on out of the bar and drive off in their pickups. "If I've seen George do it once, I must've seen him do it a dozen times," Hobbs Wilson, still impressed, recalled. "He was a regular John Wayne."

As for George, he grew to like life in the valley by the Eau Galle. The easy pace that allowed him to take off into the hills with his Winchester during hunting season whenever the mood for venison seemed to strike, the comfortable feeling that he would know the first name of just about everyone he would run into in the course of a day's work, the pride the whole town seemed genuinely to share in his son's being a big football star up at the university—all this helped to make his life and work in Elmwood very satisfying. There was only one thing that gave him pause. There was a peculiarity about living in Elmwood that he couldn't quite fathom. "George'd be hearing it all the time," his wife Doris recently explained. "Everyone would be telling him about the things they'd seen in the sky and my George would just look at them like they were crazy. 'What is wrong with these people?' he'd ask me. Of course I'd just smile and say nothing. I figured my George's time would come, too."

It was silent. That was George Wheeler's first clue. Earlier that night, when he saw the huge ball of flame coming in over a hill from the northeast, he was certain it was a 747 about to crash. It was low in the night sky—too low, that was for sure—but the pilot was doing his best to steer it away from town. George, who had been a combat flyer in World War II, silently congratulated the pilot for his skill. Still, George was very worried; this wasn't going to be good at all. He was sitting in his squad car and he had a perfect view. The 747 off in the distance kept on getting lower and lower, its flames growing brighter and brighter until its orange glow overwhelmed the night sky. As the plane moved out of town, George tried to keep pace with it in his squad car. He wanted to be nearby when it went down; there was no doubt there would be people who would be needing his help.

He drove at high speed across the empty country road, trying to catch up with it. When the plane was nearly overhead, coming in really low, just moments from impact, its flames lighting up the horizon, George pulled into a ditch. He wanted to brace himself for the crash that was at hand. It was only as the plane moved directly above him, as its light and flames illuminated his squad car with the power of a spotlight, that George Wheeler realized something was very strange. The plane was absolutely silent. There was no whirr of engines, no grinding of landing gears. *There was not a sound.* And all at once something else, equally perplexing, became apparent. The plane wasn't near to crashing. It was on a steady course.

The flying object—the officer, shaken and shaky, was no longer sure it was a plane—began to fly toward the southside flats outside of town. George, his eyes fixed on an orange light as bold as the glow from an explosion, followed. When he caught up with it, the object was, to his amazement, hovering. The land here was flat, meadows and pastures fresh with spring grass, and the light from the object lit up the countryside. It was nighttime, close to 11:00 P.M., but the sky was as bright as high noon. George could see perfectly. He was scared, trembling even, but he got out of his squad car and took it all in.

The object was a craft of some kind. It was shaped like two cereal bowls put end to end and it was hovering about 1,500 feet above the ground. It was huge—at least the size of a football field. And it didn't make a sound.

George watched it for a while. Silhouetted against the sky, it was bright and silent and motionless. He felt as if he was seeing something that wasn't real, but of course he knew it was. That scared him even more. Then, without warning, the object took off at a tremendous speed. It would be incorrect, he would later insist, to describe the craft's acceleration as quick; it was instantaneous. Even more amazing, the craft began performing what appeared to be nothing less than acrobatics. It was, George decided, putting on an air show, and he was the sole audience. The craft would start out zooming along at this truly phenomenal speed, and all of a sudden it would begin turning at sharp ninety-degree angles, one after another, as though it was being driven by a stunt pilot. Someone was joyriding in the sky above Elmwood. George could only marvel. At last, it took off, flying silently toward the west at the same impossible speed. It was gone in an instant, and the night was once more dark.

That was the first time.

The second time George Wheeler saw a UFO was almost a year to the day later, on the evening of April 22, 1976. This time there was a noise. He heard it moments before the attack.

It all began when George, out on an evening's patrol, noticed an orange glow near the quarry at Tuttle Hill. "Looks like we got a fire out there," he radioed in. "I'm going to investigate."

When he drove to the crest of County Road P, he was high up enough to have an unobstructed view. To the north, over a flat hilltop alfalfa field, there it was. "My God, it's one of those UFOs again," he shouted into the police radio. But when he started to describe the craft, he was very calm, under control.

"It's huge," he explained over the radio to Chief Helmer's wife, Gail, who was working as dispatcher that night. "Bigger than a two-story house." And he went on that it was silver-colored, perhaps 250 feet across, and that a bright orange beam glowed from its domed roof. The light was so powerful, he couldn't look straight at it. It hurt his eyes.

And just as he was describing this light, the craft started to rise. That was when he heard the loud *whooshing* noise. And, before he realized what was happening, a blue ray shot out from the craft. The ray hit the squad car.

The police radio instantly went dead. The chief's wife was yelling on the other end, "George, can you hear me? Are you all right?"

But George couldn't hear her. The car was a wreck. Its lights were out. Its points and spark plugs were ruined. And Officer Wheeler was unconscious.

David Moots, a dairy farmer, was driving the baby-sitter home when he noticed the squad car, its lights off, sitting in the middle of the road. He went over to investigate. He looked inside and saw George Wheeler sprawled across the front seat.

"George, you OK?" he asked.

The police officer didn't stir. Moots repeated his question.

This time George tried to move. He leaned forward from his seat, and then fell back. He didn't have any strength, and he looked white as milk.

"What's wrong, George?" Moots asked. He was really worried.

It took the officer some effort, but he finally managed to speak. "I've been hit. Get me to a radio." His voice, Moots noted, was shaking, full of fear.

"By a car?"

"No," George Wheeler answered very slowly and distinctly, "by one of those UFOs."

At just about the time David Moots discovered the dazed police officer, Gail Helmer was at the radio in Village Hall trying to figure out what was going on. She decided to call Paul Frederickson, a nursing home administrator, who lived just east of Tuttle Hill.

"Maybe you can look out your window, Paul, and tell me what you see. Anything unusual out there?" she asked.

It was after eleven and Frederickson was already in bed, but when a neighbor asks a favor—especially if she's the wife of the police chief—nobody in Elmwood complains too much. He went to his window.

"I saw this flaming orange object in the sky," Frederickson remembered. "It resembled a bright orange half-moon. I watched it for a full ten seconds and went back to the phone. By the time I returned to the window with my wife, the object was gone."

A few miles away, south of Tuttle Hill, Mrs. Miles Wergland was watching the eleven o'clock news on her bedroom television. Suddenly her set went black. Annoyed, she put on her slippers and trudged to the cantankerous television. She kept on pushing the on/off button. Nothing happened. And then she noticed the glorious light shining outside her bedroom window. The room was now pitch-dark but outside something was lighting up the sky—and it was moving. Its glow suddenly illuminated the bedroom. Bathed in this light, she went to the window and watched a spacecraft zoom silently across the night sky. "It was shaped like the moon, but was much brighter and colored differently," she remembered.

She watched it pass from sight, and when it was gone the television flashed back on.

The next day three people living near Tuttle Hill reported the same problem. Their televisions suddenly went off around 11:00 P.M. Despite all their efforts, they couldn't restart their sets. Then, for no reason at all, about ten minutes later their televisions started playing again. It was the darndest thing, they all agreed.

George Wheeler, meanwhile, was not in very good shape. Chief Helmer by now had driven out to Tuttle Hill and he didn't like what he found. He had known George for over a decade and in all that time, in some very tense situations, the sort of dicey barroom confrontations another man might have walked away from, George had shown his grit. He never once had backed down. But that evening his deputy was, the chief would recall with some bewilderment years later, "one hell of a scared guy."

The chief took George to Doc Springer. Dr. Frank Springer had been treating the Wheelers for over seventeen years, and he had never known George to have any serious problems. That night, though, the way Dr. Springer remembered it, "He was deeply disturbed and highly nervous." The doctor gave the officer a shot to calm his nerves, but that didn't help. Officer Wheeler was distraught. The doctor had no choice but to send him that very night to the hospital up in Menomonie.

George stayed there for three days. When he returned home, the nightmares and the headaches started. "I must have radiation poisoning," he told his wife. "That UFO zapped me." He complained so much and he truly seemed to be in so much pain that this time Doc Springer sent him to the teaching hospital up in Eau Claire.

He stayed there for eleven days. He was given all kinds of tests. The doctors found nothing wrong with him, but George found that little solace. The steady pain was still in his arms and legs, and the headaches were unbearable. He felt as if his head were about to explode. "I got radiation poisoning," he kept on repeating.

Six months after he came home from Eau Claire, George Wheeler dropped dead. His heart gave out, Doc Springer decided. Anyway, dead is dead, and Doris wasn't too anxious to have an autopsy.

George Wheeler was buried in the town cemetery. On a clear day from his grave you can see the broad, round top of Tuttle Hill looming in the distance.

Unlike George Wheeler, Carol Forster was not alone when she had her close encounter. She was with her three children. But like the police officer, the Forsters didn't enjoy the experience at all.

It all started when her husband, Bill, decided to have an adventure. They had been having their regular, weekly after-church Sunday lunch at her parents' place over in Arkansaw, Wisconsin, about twelve miles outside of town, but this afternoon her dad had found a fresh topic of conversation—his brand-new snowmobile. "You got to try it, Bill," her dad kept on insisting. "It's really something. A real adventure it is. You could use an adventure now, Bill, couldn't you?"

Bill just laughed. He was a dairy farmer; trying to make ends meet was usually adventure enough. But the more Carol's dad kept on talking, the more appealing it sounded. There was even going to be a race. A bunch of hotdogs were planning to run their snowmobiles full throttle across a couple of miles of fields and then up Martin Hill. And then *down* icy Martin Hill. Now that really would be an adventure.

"You know," Bill said after his imagination had played with the prospect for a bit, "I think I might like giving it a whirl."

There was no discussion about Carol's coming along for the ride. She was pregnant. Besides, she also had been complaining all afternoon that she still had Monday's lesson plan to prepare for her first graders at Elmwood Area. So it was settled: Carol would load the three kids into the '68 Chevy and drive on back home to her books, while Bill would go off snowmobiling. "Sorry," Bill said with a wink as Carol backed the blue Chevy out of her parents' driveway. "Guess just one of us gets to have an adventure today."

He was wrong.

. . .

Mary, the eight-year-old, was the first to see it. Since she was the oldest, she got to sit in the front seat next to her mother. The two younger kids were in the back. They had gone about ten miles across the flat Missouri Valley, white and naked in the middle of winter, the road as empty as the landscape, when Mary screamed, "Mommy, what is that?"

Carol looked to her right. A plane was coming toward the car. No, she realized, it wasn't a plane! It was too big. It was ten, twenty times the size of her car. It had a dome top. Like an overturned coffee cup, she thought; and for a split second she wondered what made her find that comparison. Then she noticed the lights. Its lights were orange and glowing. And it was heading right toward her Chevy.

The children were screaming, "Mommy, Mommy, help us."

All Carol could do was step on the accelerator. The Chevy was speeding down the flat road, but the craft was coming closer. And closer. Now the craft was directly above the car. Carol could see its landing gear. The car was bathed in an orange light. Carol was ready to panic. Mary was all orange; the harsh light from above was reflected on her little girl's face. Her baby was glowing!

"It's going to hit us," Mary screamed. And then the little girl opened the door. She desperately wanted to get out, to run.

The Chevy was doing sixty, maybe seventy, and all of a sudden the door had swung open, and Mary was about to jump out. Carol didn't stop to think; she just reacted. She grabbed her daughter, pulling her back. But as she did, the steering wheel slipped from her hand, and the car shot off the road. It went into a sharp, long skid. The kids were crying, Carol too. Everything was almost out of control. Yet somehow she got her hands back on the wheel, and the car found traction on the ice. She was able to straighten it out, keep it steady, and then she brought it back up on the road.

And the craft was still above them.

Carol, through her tears, was yelling at the children to calm down. "Stop crying," she kept on repeating, realizing how foolish she must sound since she was crying, too. Her foot was pressed hard against the gas pedal. The car was going as fast as she could make it, and when she saw the turnoff for County Road P she took it.

She drove to the first house she could find. It was three miles up the road and the craft was above her all the way. The kids wouldn't stop crying.

She pulled into Jack Baier's driveway, grabbed the kids, and ran to the front door. She was pounding on the door, yelling, crying, the kids crying, too, when Jack's teenage boy, Roger, opened up.

"Mrs. Forster, what's wrong?" Roger asked.

But Carol couldn't speak, or perhaps she figured he would think she was crazy. So she grabbed Roger by the arm, really yanked him, and led him to where her car was parked.

He got there just in time to see a bright orange light moving at a fantastic speed into the distance.

"My life has never been the same," Carol Forster said recently. "At first we thought the fear would go away. The kids slept that night all together in the living room with the lights on. They kept that up for three years. Sleeping all together. The lights always on. They were always afraid. They still have nightmares. As for me, I wouldn't go out at night without Bill. I still won't. I'm still afraid, I still don't feel safe. I still know it's out there, over Elmwood, waiting to come back. And I don't trust it. It's evil. It scares me."

It was a mystery why Tom Weber and Larry Feiler were so eager to build a UFO Landing Field; both George Wheeler and Carol Forster had been, after all, terrorized by what they were convinced were alien craft. Perhaps, however, the two men's enthusiasm was simply a product of their faith in the benevolence of the universe. Or maybe it was wishful thinking. Or, as many people in Elmwood now say, once the plans for the UFO Landing Field were announced, it quickly became clear there was too much at stake for the town to allow anyone to stop and think things through.

TWENTY-FIVE.

Mayor Feiler kept his promise. But by the time he presented the plan for the UFO Landing Field to the town board, the secret was already out. And not just in Elmwood.

Ultimately it was Tom Weber's impatience, even he will sourly admit, that was to blame. He had left his initial meeting with the mayor buoyed, glad to have found a powerful supporter, and yet he was troubled. There was a vagueness to the mayor's commitment. Weber was a straightforward man, and he expected the same sort of directness from others. As he drove back home to Chippewa Falls, he kept on mulling the meeting over in his mind and with each replay of the mayor's concluding pledge, he became convinced that a lot less had transpired than he had previously believed. Sure, the mayor had said he would bring the landing field proposal before the town board. But—and Weber's active mind now zeroed in on this increasingly important omission—the mayor had not said *when*. This week? Next week? Next month? Next year? By the time he pulled into his driveway, Weber had reached such a state of disappointment that in this aggrieved worldview any sort of betrayal was possible. Could it be, he asked himself, that the mayor had just been placating him? A guy comes in without an appointment and says there are a whole bunch of UFOs flying over your town, so why don't we just build them an airport and invite them to come on down and visit—hell, Weber was beginning to suspect, you just might want to humor the guy. Let him say his piece and then be on his way. Maybe, he told himself, that was precisely what Mayor Feiler had done.

In that anxious manner it came to Tom Weber that it would be a mistake to count solely on Larry Feiler. A shrewder strategy, he plotted as he lay awake in bed that evening, would be to push on immediately with his great plan. He would set out on his own to attract supporters. Why, he began to ponder as if the idea was suddenly real for the first time, the purchase of the land—two square miles!—was going to take a lot of money. More intimidating, he couldn't even begin to estimate accurately the millions the construction of the tarmac, the illuminated welcoming display, and the scientific facilities were going to cost. Yet as he played with the possible numbers in his mind, as he decided that the fifty-million-dollar estimate he had shared with Feiler still seemed to his businessman's brain to be the logical ballpark figure, his resolve was strengthened. He would move on without waiting for the mayor. If Feiler was truly interested in helping, terrific. There would be time for all of Elmwood to come on board. But he would not wait. The landing field was too important. Too much, he was certain, was at stake. He would act now.

The next day Tom Weber placed an advertisement in the *Eau Claire Leader-Telegram*. It was a small ad, laid out in a four-by-four-inch box. In not more than a hundred words, he announced the plan to build a landing field for UFOs in Elmwood, Wisconsin. Would anyone interested in helping out in this project—a positive, scientific enterprise, he emphasized—please contact him. He signed his name and gave his business phone.

The ad appeared on a Tuesday. By the weekend, the story had not simply leaked, it had exploded. It seemed the small ad had managed to attract some curious inquiries from a reporter on the *Leader-Telegram*, Chuck Rupnow. He wrote a straight-faced, newsy account that began on the front page. UFO GROUP PLANS LANDING SITE FOR ALIENS was the headline, while UFO GROUP PLANS INTERNATIONAL FUND RAISING was the optimistic bold-faced head on the jump. The story prompted Rob Kreibich, a reporter for WEAU-TV in Eau Claire, to do an interview with Weber. It was a rather flattering portrayal; but then again, Kreibich was a believer. And it was this interview that caught the attention of the local Associated Press bureau. A story headed "UFO Landing Field Proposed in Wisconsin Town" went out on the AP wire. And around the world.

It was after that, Larry Feiler remembered, when "the phones started ringing in Elmwood. It got to the point where they wouldn't stop. Why, we started to get calls from as far away as Australia, British Columbia, and South Africa. I was getting calls from Geraldo, from *PM* magazine. One from Pierre Salinger."

As it turned out, many of these people were calling the mayor of Elmwood for one simple reason—they couldn't find Tom Weber.

. . .

Tom Weber wasn't in Chippewa Falls, he was in Harlingen, Texas. He had been told the *Leader-Telegram* story wouldn't be running for another two weeks, so he had decided to take advantage of the lull. He had gone south, to the home of his wife's grandmother, not to find UFOs, but simply to find some sun. He might just as well have been looking for alien spacecraft. It had rained for four straight days; and a soaked Texas prairie, both he and his wife had rapidly come around to discover, was just as depressing as a snowbanked Wisconsin plain. Maybe even more so.

All in all, it had been that kind of vacation. He and Colleen, full of great expectations, had left the kids back home with a baby-sitter. They had hoped that this trip, just the two of them, a second honeymoon of sorts, would be rejuvenating, a chance for them to relax and work out some of the friction caused by their differences. Actually, there was only one sharp thorn in their marriage—Tom's "damn science." His tendency to expound his theories on the electromagnetic physics propelling extraterrestrial vehicles was taking its toll. "Truth was," Colleen would complain months later, her voice still sharp with exasperation, "it wasn't that I didn't understand what Tom was jawing about. I just didn't care. I was sick of it." And his new plan for a landing field for these ships was, she remained convinced, more folly. "What about his business, his family, I kept asking him," Colleen remembered. "People were going to look at us as if we were a little weird. I mean, airports for Martians? What would you think?"

Sitting in the house watching the gunmetal gray Texas rain didn't help Colleen's enthusiasm for the project. Tom tried to persuade her, but she wouldn't budge. A volatile situation, Tom was worrying, was about to become dangerous. Then he—and maybe the kitchen china—was saved by a phone call.

It was early Sunday morning and the Webers were still in bed when the call came from their baby-sitter in Wisconsin. "The phone here's going crazy," she explained, sounding rather desperate. "Reporters are calling up from everywhere, even from New York, asking about Mr. Weber and some sort of alien airport." When she had disclosed that Mr. Weber was out of town, the sitter recounted to the startled couple who were now listening on extension phones, the more aggressive newshounds refused to be put off—they interviewed her. Live! On radio! "Well, I don't know much about this alien airport or whatever you call it, but Mr. Weber, yeah, I know him. He's a good guy. I can tell you that." There you have it, folks, an exclusive interview with the baby-sitter of the man who plans to have aliens land in Wisconsin!

The call from the frantic baby-sitter had three immediate consequences. First, it was excuse enough for the Webers to decide to cancel the rest of their stay in Texas. Wisdom (not to mention the weather) suggested that they had better go home and rescue their children from the harried baby-sitter's clutches. Second, it convinced Tom Weber that the time had come to hold formal meetings with his supporters, who, he was elated to find out, seemed to be many and eager. And third (and perhaps most important of all for Tom Weber; four soggy days with a disenchanted wife can do a lot to narrow your historical perspective, he was learning), it helped persuade Colleen that maybe her husband's idea was not so crazy after all. If people from all around the country—even Geraldo!—wanted to talk with him, then she, too, was ready to believe he really did have something significant to say.

The first meeting of the UFO Site Center Corporation was held on a Wednesday evening in the small conference room of the Chippewa Falls library. It was a select group. A lot of interested people had contacted Tom Weber in the busy weeks following the appearance of the *Leader-Telegram* advertisement, but not all of them were invited. He had, his patience soon rubbed to a shiny raw by the day-and-night stream of outlandish calls, peremptorily eliminated the man who had wanted the concession to sell flying saucer key chains at the landing field, as well as the man who could supply (for a suitable fee) maps of the universe with X marking the spot of planet Earth so that when the aliens landed and took earthlings as passengers, the cosmic pilots would be able to find their way back to Elmwood. And he had hung up with a suddenness that he, in retrospect, found both embarrassing and disconcerting on the man who had photographs of buildings on Mars, pictures he would be willing to give to Weber in exchange for a position of responsibility with the Site Center. The people invited to this meeting were the ones Weber, becoming comfortable with the prospects of great responsibility, was persuaded he could count on. They were, he boasted, serious people. They would be the men and women with whom, he was certain, he would one day stand in the UFO Site Center control tower as the first ship came in for a landing. "My cadre," he called them.

There were about a dozen at the meeting, all locals and all believers. There was, for example, Dave Martinek of Eau Claire, who had become fascinated with flying saucers while in high school; he saw the landing field as part of Earth's "obligation." "What if there is another life-form out there waiting for this Site Center?" he, genuinely concerned, had suggested. And there was Lee Horne, the Chippewa Falls nursing home administrator, who became the group's treasurer. She had never seen a UFO, but she was already prepared for the inevitable day when she would spot one: "Take me with you," she planned to demand. The others were fueled by a similar sense of commitment. "We were aware from the start we were doing something historic," said Lee Horne. "If we were successful, it was going to be the greatest news in the history of the Earth since the coming of Jesus."

It was because of this motivating sense of history and duty that the meetings of UFO Site Center Corporation became a weekly Wednesday night event. They were hectic sessions. There was so much to plan, so much to decide. And of course, with something of such momentous consequence at stake, and with so many believers who sincerely felt their insights, intuitive as well as scientific, were the keys to understanding the mysteries of the cosmos, there was a lot of debate. Things could get pretty heated at the Chippewa Falls library on a Wednesday night.

There were, just to cite one acrimonious discussion, many different views about what form the welcoming illustration should take. One loud voice had a very meticulously reasoned theory involving geometry, and precisely how squares, triangles, and circles could be put together in a way any alien would be certain to read as an invitation. Another approach for attracting the extraterrestrials involved staging a musical program, actually more an extravaganza complete with costumes like a Broadway show rather than a serenade. While a third suggestion, and one that won a few supporters, was to have a larger-than-life illuminated depiction of a man and a woman copulating; what, it was asked, could be more inviting and more indicative of our peaceful intentions than such a primal scene? The idea, though, fell by the wayside after it was pointed out that some of the residents of Elmwood might not be too happy about having a naked couple, each of them stretching fifteen feet from head to toe, going at it, day in, day out, in one of the town's alfalfa fields. So in the end it was decided to have what Tom Weber called "a simple, logical, straightforward illustration."

With Weber's guidance, a local artist had come up with a sketch of a man preparing to shake hands with an alien. The man didn't resemble the bearded, big-domed Weber or any of the heavy-bellied Elmwood farmers or even the town's slicker political leader. Rather, he had the archetypal Aryan features of Barbie's boyfriend Ken. And, in the interests of convenience (as well as modesty), he was dressed in what appeared to be a black jumpsuit. The alien had a similar outfit, Ken's hands minus two of the fingers, two solid feet, and a head the size of a watermelon.

Still, there was some concern after a mock-up of the pair of figures was presented at one Wednesday meeting. Is it possible, someone asked, that a handshake might not mean the same in the Andromeda Galaxy as it does in Chippewa Falls? Suppose, it was suggested, a handshake is a vulgar gesture to an alien? That we're illuminating a cosmic "Screw you" to the first visitors from another world? Or perhaps a handshake could mean "Let's fight"? Or even "Good-bye?" But Tom Weber was undeterred. "If they're smart enough to get here, they're smart enough to figure everything else out," he ruled. The handshake would remain.

And when the cadre wasn't debating matters of philosophy, it was busy fund-raising. It was a nonprofit corporation and adamant about neither seeking nor accepting government funds. The only hope, then, was private citizens and his troops embarked on a variety of schemes that left Tom Weber, a proud man, feeling a little shamefaced as well as more than a little surprised by his own latent aggressiveness. With Weber cheering them on, they hawked sky-blue T-shirts emblazoned with the convivial suggestion LET'S MEET, and available in a "poly/cotton blend" in small, medium, and large sizes for a donation of twelve dollars, shipping included. They distributed flyers that explained the significance of their landing field and requested pledges of either monthly or onetime contributions. And they set out systematically to solicit just about anyone they had reason to suspect of believing in UFOs. It didn't matter if the potential donor had made his comments about flying saucers late one night in the Sandbar, or in the pages of the *Enquirer,* or even in a stray aside to Johnny Carson with the whole nation watching. The Site Center Corporation was intent on tracking them down and asking for funds. And it worked. They were able to get a hundred dollars from the Elmwood library, five thousand from a retired Elmwood resident who, in his younger days, had dreamed up a lot of inventive farm machinery and now held the lucrative patents; and they were able to get a meeting with Muhammad Ali—almost.

It was a tribute to Tom Weber's doggedness that, after reading that Ali had once had a sighting, he was able to track down the former heavyweight champion. And not only was he able to find out where Ali lived, Weber told his cadre with unrestrained excitement, but he was also successful in getting Ali on the phone. Ali himself! Not some manager or gofer. And the great Ali had said that he was interested. Sure, Ali had agreed, come on down to see me in Chicago and I bet we can work something out. I'd like to hear some more about what you're doing. Maybe I can get involved. And so with visions of an endowment dancing in his head, with Rob Kreibich of WEAU-TV in tow to record the historic moment, Tom Weber set out for his meeting with Muhammad Ali. Except when he got there, Ali was not at home. The champ had to leave for Indonesia, an unexpected trip, Ali's wife apologized. The long ride home to Chippewa Falls was a rough one for Tom Weber.

But he bounced back. There were other celebrities out there, he told himself. Maybe Steven Spielberg or Donald Trump would return one of his calls. In the meantime, money, a bit, was coming in. So the Site Center rented a two-room suite just down the hall from an accounting firm and a psychiatric clinic in a small brick office building in Chippewa Falls, bought a folding table and four metal chairs, and began to concentrate full-time on the serious business of planning the landing field that would change the world.

Forty-five miles to the northeast, over in Elmwood, Larry Feiler and the town board were also busy making plans. Only the mood here was not at all like the somber plotting going on in the Site Center's dim two-room suite. The Elmwood city fathers were nothing less than gleeful. What an opportunity the UFO Landing Field had turned out to be! Elmwood—quiet, isolated, 991-person Elmwood—was about to become a boomtown.

Now that the news had been announced, now that the phones were ringing off the hook, the prospect of Tom Weber's building a $50 million UFO airstrip in a town where the annual budget was $283,411 had created a frenzy of eager and greedy anticipation. "This town is looking at considerable impact in terms of jobs and the local economy," Mayor Feiler predicted. He estimated that as many as a hundred scientists and professionals would be working at the Site Center—and, though it went tactfully unsaid, all living and spending in Elmwood. "We're just thrilled with it," effused Wayne Nohelty, the town's banker, trying hard, it might be imagined, not to count his money too soon.

But even though the landing field and its two fifteen-foot handshaking welcomers were still, literally, on the drawing board, there was real cause for immediate celebration. "This town is digging in for what may be one of the longest tourist seasons in its history," Mayor Feiler had announced. Everyone in town, it seemed, had great hopes for this year's UFO Days celebration.

Ten years earlier, back in 1978, the town had decided that its annual Stars and Stripes Festival was not working out as had been anticipated. While other nearby Wisconsin towns with their Cucumber Days or Mosquito Festivals were attracting their share of free-spending tourists, Elmwood wasn't getting very many visitors. Only a handful of tourists had been willing to drive out of their way to see another American Legion parade led by another bunch of ragtag, flag-carrying veterans. Desperate, the Elmwood Community Club decided to sponsor a contest to rename the event. The five-dollar grand prize went to a farmer who had suggested a theme based on the town's major preoccupation—UFOs.

UFO Days, always celebrated during the last weekend in July, were an immediate success. The Community Club and local merchants could count on as many as three hundred visitors clogging up Elmwood's Main Street on those two festive days. But crowds like that were before all the hoopla about the UFO Landing Field. Now, prospects were looking better. A lot better.

Jeff Martina, executive director of the Pierce County Economic

Development Corporation, announced at a planning session of the UFO Days committee that "there is no way any of our communities could have afforded to put Pierce County in the press to this extent . . . The publicity has got to be a boost." While Mayor Feiler went around town telling a joke. "Where's Madison, Wisconsin?" he would ask. And then before anyone could answer, he would blurt out, "Five hours south of Elmwood."

The town of Elmwood, he would insist to anyone who was willing to listen, was serious about its support for the UFO Landing Field. In fact, the town, on the mayor's recommendation, had even provided Tom Weber and Site Center with the use of the Village Hall auditorium during UFO Days. And, further proof of his personal commitment, Mayor Feiler was making his relationship with the landing field part of his reelection campaign. "A vote for Feiler is a vote for the Site Center" went the succinct slogan.

And while it didn't make for a catchy chant, there is no denying that it was a shrewd bit of politics. The Site Center offered the town the prospects of an appealing, mutually beneficial partnership. Without the ground having even been broken, there were already rewards: Elmwood was expecting five, maybe ten thousand visitors—more than ten times the size of the town's entire population—to visit during UFO Days this year. Now, who in Elmwood could be opposed to that kind of blue-chip opportunity?

Except, one man was.

TWENTY-SIX.

On the Sabbath in Elmwood, the pews in the United Methodist Church always began to fill early. The hatted ladies, their broad backs ramrod straight, sat shoulder to shoulder with their somber, suited husbands. There was a murmur of conversation, a soft undercurrent of noise as friends caught up on what had come to pass in their valley in the last week. But all this dissolved into a sudden hasty quiet, like an almost audible intake of breath, when the Reverend James Thunstrom approached the pulpit.

As the minister stood there, looking down at his congregation, he seemed as tall and as tapered as the church's white steeple. He took his time before speaking, searching at this last moment, one might suspect, for the steel of inspiration. And, in turn, his congregants, silent, fixed rigidly to their seats, regarded him with a rapt, anticipatory attention. For while it would be sacrilege to make light of the Elmwood Methodists' devotion to prayer and the worship of their God, it would also be naive to say that for many James Thunstrom was anything less than the high point of the service. He was a youthful man, not quite forty, with a cowlick in his sandy brown hair and soft, moist eyes, but once he began to speak all those boyish qualities seemed to recede. He could put on quite a show. He, when truly moved, reached down deep into some pit of wisdom and anger trapped in his soul, and the words came out singed with fire and brimstone. The Message boomed forth from him like thunderbolts. And his congregation, awed, mesmerized, and in the end inspired, loved it.

And so it came to pass on a warm, bright Sunday morning in July 1988 that the Reverend James Thunstrom stepped forward and began with the reading. The text, as usual, was from the Old Testament. It was Exodus, chapter 32:

"And when the people saw that Moses delayed to come down from the mount, the people gathered themselves together unto Aaron, and said unto him: 'Up, make us a god who shall go before us. . . .' "

The minister's voice that morning was, as always, loud and rich; it reached up toward the arched gables, filling every corner of the wooden church. It was his practice to lean in close to the text as he read, his eyes almost parallel to the Book. And his lanky form angled above the pulpit like a carved figurehead on the prow of a fast-moving ship, he continued:

"And Aaron said unto them: 'Break off the golden rings, which are in the ears of your wives, of your sons, and of your daughters, and bring them unto me. . . .' And he received it at their hand, and fashioned it with a graving tool, and made it a molten calf; and they said: 'This is thy god. . . .'

". . . and the people sat down to eat and to drink, and rose up to make merry."

When he was done, he closed the Book with a portentous thud. There was, it would be recalled, not a rustle from the congregation, not a stray cough, not a wandering eye, as the minister, once more erect behind the pulpit, began his sermon. It was a hot one.

Later, over Sunday supper and even for months afterward, some congregants would attempt to explain away what had ensued by suggesting that their minister was simply giving vent to the anger and rage that had been swelling up throughout the reading of the text; what he could not direct at the mercurial children of Israel, sinners quick to abandon their God for a false one, he hurled instead at the stunned children of Elmwood. But others, try as they might, could find no comfort in easy excuses. James Thunstrom's exhortation, all of his stern, condemning fury, was, they woefully acknowledged, meant for no one else but his wayward congregants.

His Message that memorable Sunday had, obliquely, begun with a declaration. I will put no gods before the Lord my God, he solemnly proclaimed, some in the congregation later recollected. (And since no one, of course, was taking notes that morning and the minister, typically inflexible, will not even consider discussing this particular sermon, these still vivid memories will have to suffice.) The congregation was still puzzling over where that loudly voiced vow might be leading when the minister, with a sudden thunder that some found more frightening than nature's own, made another pious oath. He, for one, refused to bow down to a golden calf. And with these pledges publicly affirmed, with his right arm now outstretched, one long finger pointed straight at his apprehensive flock, he set off to admonish the Methodists of Elmwood. He chastised them for the faithlessness they had already exhibited; and he pilloried them for the large sins in defiance of the First Commandment they, with so much enthusiasm, were at this very moment preparing to commit.

It was a lecture in human fallibility, and the minister pursued two routes to reach the same spiritual destination. The first was the high road of reason. The UFO Landing Field, he contended, was absurd. "The whole thing lacks logic and any depth of real thinking about it," he would later tell a local reporter. But at the time, he simply was cutting. Think about it, he asked his congregants, why should flying saucers want to land in Elmwood? And if they wanted to visit, would they really need an invitation? Who's to stop them? How can our town, he asked, condone and support such a ludicrous endeavor? Have we lost our senses?

The second argument he offered took an even higher road—that of Christian doctrine. This was the crux of his attack (and, no doubt, the source of his apocalyptic anger) and he was imperious and unforgiving. "It is a misguided effort to seek answers or help from spiritual forces other than God," the minister later curtly explained when someone asked what he found so troubling. But that Sunday morning to his flock he was, by many accounts, more exhortatory.

We are worshiping flying saucers, when we should be worshiping God, he lashed out. We are waiting to see false signs, when we should be admiring His real and truly wondrous works that are all around us. Our town, he went on without pause, his pitch and cadence solid, has lost its faith, its very center. We have turned from God and are worshiping a golden calf. Woe unto us, he cried as though in real pain, which he no doubt was. Woe unto Elmwood.

But it gets worse, he said. Our sins breed other sins. They multiply. And now he paused, but whether it was for drama or simply to catch his breath was never made clear. When he continued his voice had a mocking, sinister twist. It was the voice, some of the congregants decided, of the Serpent that begged Eve to take just one bite from the apple.

And it doesn't end, he said, with you meekly acknowledging this false god, does it? No, you must celebrate it. You must indulge yourselves. You must eat, drink, and make merry in adoration of its perverse authority. Like the faithless Israelites, you prepare to dance around the calf. Well, I will promise you this. The fate of your souls will be nothing less than the divine judgment that was rendered on the Israelites. You will be condemned. Woe unto them. And woe unto you who celebrate the pagan bacchanal that is our UFO Days.

Yet why, the minister demanded, do so many good and worthy citizens of our town suddenly want to abandon their God and worship this pretender? The answer, he told his audience, is a particularly venal one—Mammon. Woe unto them who will sell their eternal souls for the ephemeral profit that can be made from two days of pagan celebration.

He concluded—and by now a half hour or so had gone by; but who would have dared to sneak a glance at his watch to check?—in a softer, almost melodic voice. Yet his final words offered not even the hope of a compromise. He and his family would not stay in Elmwood for the UFO Days weekend. He expected all good Christians to do the same.

And though the Reverend James Thunstrom found no need to search for other theological fuel to stoke his fire that Sunday morning, it could have been easily provided. Elmwood's Methodist minister was not alone among Christian thinkers in his adamant rejection of contact with extraterrestrial life.

In fact, while he launched his attack as a diatribe against a vague, misguided paganism, other men of the cloth were more direct. The devil himself was out there. Aliens would have to be, these doctrinaire Christians realized, evil. Nothing would convince them otherwise. It was a simple matter of dogma. Redemption was a gift intended only for humanity. Life on other worlds, therefore, was damned.

In the Gospel, they found their armory of logic. It provided the spiritual strength to support their fears. It taught that God is omnipresent and so is individual evil. The corruption of created beings that began with Adam and Eve was a universal state, one that predated the Immaculate Conception and the miraculous birth of Christ. And without Christ's sacrifice, salvation—the ability to choose between good or evil given by grace—was impossible.

But—and this was the pillar of their eschatology—there was only one Jesus—*our* Savior. Christ had died on Earth, and he had not died for *their* sins. To suggest otherwise was not only wrong, it was also blasphemous. Imagine, a multitude of planets with a multitude of crucified Christs. Only how could there be wooden crosses in worlds where there were no apple trees; or figs that the Son of God could curse on planets where no figs grew; or virgin births of beings in galaxies where the reproduction of the inhabitants was not begun by copulation? The images from the Gospel could not logically be extended to other worlds.

And what could not be, was not meant to be. They—whoever was out there—were not meant to be saved. Redemption had most assuredly been denied to them. But why? There could only be one reason. Life on other worlds must be inherently evil. It was a fearful thought, and it provoked a daunting corollary. Why is the universe so enormous? Its very vastness, C. S. Lewis, the Anglican lay theologian, reasoned, has, like all things, a divine purpose: to "prevent the spiritual infection of a fallen species from spreading."

There were not angels inhabiting the heavens, but demons. A threat to mankind and Christianity. And now Elmwood was preparing to welcome Satan; and, for two days in July, to bow down to him.

Woe unto the people of Elmwood, the Reverend James Thunstrom had said.

But theology splits many ways. Across town from the Methodists, over at St. Peter's Lutheran and at Sacred Heart, the congregations heard not a discouraging word about Elmwood's great endeavors. At St. Peter's, perhaps, that was to be expected. The Elmwood Lutherans had a tradition of refusing to mix their spiritual and political demons in the same sanctuary; and the Site Center and UFO Days had become, especially after the way the town board and the mayor were promoting them, political issues. At Sacred Heart there was no such policy of reserve; Father Thomas could be very outspoken. However, the only comment—if it could be called that—was made rather quietly. On the parish house community bulletin board someone—it was never discovered who—had tacked a white three-by-five card. On it was a quotation, faultlessly typed, that was attributed to Father Theodore M. Hesburgh, the former president of Notre Dame. It read, in part:

"It is precisely because I believe theologically that there is a being called God, and that He is infinite in intelligence, freedom, and power, that I cannot take it upon myself *to limit what He might have done.*" It was a community space, and anyone could tack up whatever notice he saw fit. Its authority was certainly not ex cathedra. Still, it would remain on the bulletin board for over a year and many of the town's Catholics came to believe that must mean something.

The Reverend James Thunstrom, as could be expected, kept his promise. He and his family left town during the UFO Days celebration. Whether any of his congregants also did so, is, his fellow Methodists insisted, "a private matter." But, regardless, perhaps it was just as well that the minister, his family, and maybe a few others fled from Elmwood for those two days in July. Because, as things worked out, the town needed every inch of elbow room it could come up with. Downtown Elmwood was a mob scene.

TWENTY-SEVEN.

By nine in the morning on that last Saturday in July, the road into town was jammed. Cars, pickup trucks, vans, and hulking, wide-bodied campers were backed up bumper to bumper along route 128. A man in blue denim overalls, tall and as skinny as a fishing rod, was standing up in the open rear bed of one of the pickups, and as the truck sat there locked in traffic, not moving an inch, he began to play his guitar. He sang (for reasons only he knew best) "If I Had a Hammer." A few motorists began to sing along. And then others joined in. Their loud voices lifted up above the valley. And when he was done, they cheered him and they cheered themselves. Car horns honked with applause. It was a very jolly, communal procession that was making its way to Elmwood. But perhaps that was to be expected. They were pilgrims, after all, and it must have been very heartening to see proof all around them that so many shared the same special faith.

On Main Street, though, Chief Helmer was in a state. He was patrolling a strategic downtown corner—near the stop sign and across from the welcoming UFO DAYS '88 banner—and he also was coming around to the unpleasant realization that he had goofed. Naturally, he had anticipated a larger turnout this year; that was to be expected after the ruckus the landing field had kicked up. But never had he dreamed of anything like this! Forget flying saucers, he felt like shouting. *This*—a sea of people in little Elmwood—was truly incredible. And troubling. He had recruited five special deputies for the weekend, but he couldn't realistically expect them, farm boys with badges, to handle the situation if it really got out of hand. It would be wiser to call the state police, get some troopers to come on down just in case.

But, typically, he decided there was no rush. He would give things around town a quick once-over before heading to his office phone in Village Hall. And as he made his meandering tour of downtown Elmwood, a new concern began tugging at him. Where were all these people going to sleep? There wasn't a single motel in Elmwood. Not even anything that could pass for a guest house. Someone's going to have to tell these people they'll need to head off to Durand, or Menomonie, or maybe even all the way to Ellsworth for the night. Or so he had thought at first. Yet as he moved on, he discovered he had underestimated the resourcefulness (and perhaps business acumen) of his neighbors. At house after house, he saw signs firmly planted in the middle of postage-stamp-sized lawns. The signs read, CAMPING PRIVILEGES AVAILABLE. INQUIRE WITHIN.

That gave him a smile, but as he crossed Shaw and headed back toward Main, his good humor collapsed. It was a zoo. Just getting across Main Street was an adventure. A bizarre invading army had taken over his hometown. Forget the landing field, he was near to shouting. The aliens had already landed, and they were everywhere.

Families—booming husbands, doughy wives, and screaming, sniffling brats—marched two abreast like ducks. A man in Bermuda shorts, his bottom nearly as broad as a cowcatcher, was maniacally taking photographs. He even had the gall to ask the chief if please, sir, he'd pose; hell, what'd he think the police chief of Elmwood was, a cigar-store Indian? And, a new sharp worry, there was a pack of scraggly hippies, long-haired boys and giggling, braless (*definitely*) girls, in ripped jeans, their pockets stuffed with God knows what manner of illegal drugs, sashaying down Main Street like they owned it. And some of them were barefoot, no less.

Where had all these people come from? He saw cars and vans with Texas, Oklahoma, even California license plates. And later he would read about the guy from the Philippines who had come all the way to Elmwood. Which, the chief would say, was "really something." But then again, so was the purpose of Bernardo S. Ceguerra's journey. He was on a mission to locate a boy in the "Wisconsin-Indiana area" who had been dropped off by a UFO in 1979. He knew this to be a fact since, on three separate occasions, aliens had approached him in his native Philippines and repeated the identical story verbatim.

Lighten up, the chief admonished himself as he continued his tour. It's only till Sunday night. Elmwood would survive. And, he couldn't help but notice, a lot of his fellow citizens were certainly taking the crowds in stride. They had set up a row of concession stands on Main Street and he wondered if they were giving things away for free, the way tourists were lined up. But, he quickly found out, nobody from Elmwood was giving anything away that weekend. His friends and neighbors were selling "Lunar Linens" (which had a remarkable resemblance to the Homer Hankies that the local Twins' fans were always waving) and "UFO Burgers" (which looked suspiciously like the hamburgers at the Sandbar) and "Flying Saucer Pizza" (which, too, seemed to have a lot in common with the more prosaic slices usually available at the Shack). And all this food and merchandise was being sold at prices that, he couldn't help but think, came dangerously close to being larcenous.

So much for quiet, neighborly Elmwood, he would remember deciding as he, at last, reached Village Hall. And there to greet him, standing out front and taking it all in, was Mayor Feiler.

"What do you think?" the mayor asked.

And before the chief of police could even let out a sigh, the mayor, smiling like a man who had just won the lottery, gazed out at the crowded, hectic streets, and bellowed, "Beautiful, ain't it?"

Yet, by the time the sun had set on Sunday night, some people couldn't help but be disappointed by the UFO Days festivities. Carol Sage was just one of many discouraged visitors. She had made her way to Elmwood from her home in Boston, admittedly "not quite knowing what to expect." But still, she said with apparent gloom, she had been anticipating "something more" than what she had encountered. She had traveled halfway across the country to take part in a New Age festival, a cosmic Be-in. Not a small-town county fair.

Such gripes were fighting words to the Elmwood Community Club. Yes, its members would admit with disarming candor, they did have to cancel at the last minute the bus tours of "Famous Elmwood Saucer Sighting Sites." That couldn't be helped; corn had been planted in the fields and some of the farmers were, understandably, not eager to have their crops trampled by a horde of tourists. But, even without the bus tours, the celebration, they hotly asserted, did have a

"UFO-y feel to it." And to prove their point they shared the official "lineup of events":

—A UFO medallion hunt beginning at 7:00 P.M., Friday.
—Sidewalk sales offered by Elmwood merchants.
—A splatter ball shooting contest.
—A basketball camp and softball tournament.
—A pancake breakfast beginning at 8:00 A.M., Saturday.
—A double-elimination volleyball tournament all day Saturday.
—A 10:00 A.M. Kiddie Parade on Saturday followed by children's activities conducted by Bob Afdahl and members of his Elmwood wrestling squad.
—A horseshoe and cow chip throwing contests.
—A paper plate drop beginning at 1:00 P.M., Saturday.
—An amateur talent contest beginning at 6:30 P.M., Saturday.
—The UFO Grand Parade beginning at 2 P.M., Sunday.

And, members of the Community Club pointed out, their voices loud with indignation, that list didn't even begin to reflect just how "UFO-y" the events really were. Take, one of the club directors suggested quite randomly, the cow chip throwing contest. It wasn't your ordinary chip throw. When the competition took place to see who could hurl a piece of dried dung the farthest, the judges called it "The Flying Saucer Event."

Throughout the weekend, Tom Weber was stoically determined, regardless of whether the activities were UFO-y or not, to remain aloof from the hoopla. He stayed, along with his wife and a few loyal supporters, in the relative isolation of the second-floor Village Hall auditorium. "Puts us above the fray," someone, perhaps Lee Horne, the Site Center treasurer, had joked. To which Weber, never one to treat things lightly, agreed, but in his own way. "We're here because we want to address the serious aspects," he reminded all who asked. Besides, he was confident that anyone who was sincerely interested in his UFO Landing Field and the future it promised would take the trouble to come in from the street and climb the two flights.

And he was right. For the entire weekend, the auditorium was jammed. People listened to Weber's plan, studied a mock-up of the two handshaking intergalactic greeters that would welcome the arriving aliens, and bought Site Center T-shirts, many of the visitors putting on the shirts right then and there. A cigar box with the word CONTRIBUTIONS labeled on its lid was filled by the end of the first day. It was very exciting.

Not everyone, though, was completely enthusiastic. Susu Jeffrey, for example, thought there really was no need to build a fifty-million-dollar landing field in Elmwood. She had, she told a skeptical Weber, already built one next to her home in Minneapolis.

Her field, she explained, was eighteen feet in diameter and had a twelve-inch-thick base of pebbles she had salvaged from the Mississippi River. All told, she estimated, she had spread at least three tons of pebbles to protect the ground from the hot blast of a UFO's takeoff or landing. "It's scorch-proof," she bragged.

Weber, though, was unimpressed. "Like comparing a child's toy to the real thing," he scoffed.

Anyway, there was no point in getting into a debate. There were too many others to meet and to persuade. Throughout the entire weekend, there was always a long, steady line of inquisitive visitors snaking its way up the narrow Village Hall stairwell to the auditorium And among the curious were two men who claimed to be NASA engineers, which, when he found out, surprised and thrilled Weber some. But, no doubt, he would have been even more surprised (and conceivably more thrillled) if he had known that their impressive occupations were just a bit of cover, and that the two men were in reality CIA operatives gathering information for the UFO Working Group.

TWENTY-EIGHT.

Each session in the Tank had a character of its own. There were meetings of the UFO Working Group where cosmic mysteries swirled about the soundproof room and the infectious excitement of being part of the hunt of a lifetime was at hand. Other days, the mood was more rigorous. Science, its strict and countless facts and formulas, set the tone; and at least a couple of the participants found they were soon lost, pushed out way beyond their depths of knowledge as the parade of nuclear engineers, physicists, and astronomers droned on. While on still other occasions, the meetings would be sheer bureaucracy, detailed accountings of all the assorted nuts and bolts that would need to be carefully tightened before a specific investigation could go operational. But, it was generally agreed, the discussion of the "Elmwood affair" was the most frustrating session in the committee's history.

Colonel Phillips summoned the meeting to order by throwing a T-shirt on the conference table. It was sky blue and the words UFO SITE CENTER were in bold block letters just below the neck; while toward its hem, in a more cheery script, was the suggestion, LET'S MEET. In the middle of this circle of copy stood the landing field's official greeters—the jumpsuited, big-headed alien and his squeaky clean Homo sapiens buddy. They were, like the forever panting lovers on Keats's urn, near to but never quite shaking hands. A few of the people in the room studied the shirt and had a small laugh, which they might have expected was the required response.

They were mistaken. This was no laughing matter for Colonel Phillips. He was exasperated to the point of anger. He had produced the T-shirt, he explained testily, because it was the only tangible product the Working Group had to show for all the time and energy that had gone into its investigation of the Eau Galle River Valley sightings.

A question was asked: "So the people in that town were simply making things up?"

The colonel said he didn't think that was the case at all. After reading the CIA's report, he was convinced the residents of Elmwood sincerely believed they saw *something.*

"But now we know what they really saw, right?" the questioner persisted.

The colonel's answer took some time. He explained that the field teams had interviewed (using cover identities, of course; but that went unsaid) many of the townspeople who had claimed to have seen space ships; that the ground areas near where the "craft" were sighted were not only visually inspected, but also soil and geological analyses of the terrain had been undertaken; that there had been an effort to coordinate the exact dates of an alleged appearance of a flying saucer over the skies of Elmwood with any Flash abnormalities either NORAD or SAC might have in their records (the colonel, however, volunteered that this was less than a scientific exercise; memories in Elmwood of exact days and times had grown a little fuzzy with the passing of years); and that there had even been a study (again under arm's-length cover) of the medical records of certain Elmwood citizens who had had encounters with the "ships."

As the colonel made his report, his weary tone, the deeply disconsolate set to his face, was clue enough to some in the room as to what all these investigative efforts had produced. But the colonel, forever dutiful, seemed to take some comfort in reading from a thick, previously prepared script. He recited a long list of all the possible rational explanations for the sightings. It was a list that included such candidates as weather balloons, experimental Air Force jets, low-flying satellites, and even the old standby of swamp gas. Each one of these explanations, the colonel stated, had proved inapplicable to the situation in Elmwood.

Finally, he put down his pages and looked directly at the members of the UFO Working Group. "Gentlemen," Colonel Phillips announced rather forlornly, "we just don't know what's in the skies over Elmwood."

TWENTY-NINE.

Six months after the UFO Working Group had met to discuss Elmwood, I saw my first extraterrestrial. It was a chilly, gray California day and I had driven to the NASA Ames Research Center at the Moffett Field Naval Base to discuss SETI. Jill Tartar, one of the project scientists, after telling me that she "grew up in the generation when there were Saturday morning Flash Gordon shows" and therefore "never, ever, *ever* considered the possibility that we were alone," offhandedly suggested that as long as I was there, I might also want to view NASA's collection of extraterrestrials. I was directed to a long hangar, and when I walked through the door, I was confronted with a variety of creatures from other worlds. Or at least the drawings and three-dimensional models I began to study were what government scientists believe extraterrestrials could very well look like.

Exobiology is the name scientists have given to the study of lifeforms as they may appear outside Earth's environment. It is the only branch of biology, John Billingham, the director of NASA's exobiology program, has pointed out, that has the audacity to attempt to study the life processes of an organism without having proof that the species even exists. It is a field where scientists consider what they know about molecular, organismic, and ecosystemic evolution on this planet, and then, bold as gamblers, they attempt to make, as C. Owen Lovejoy of Reading University, in England, explained it, "quite reasonable inferences" about what life might be like elsewhere in the universe. It is part science, and part speculation. And in this brightly lit exobiology laboratory at Moffett Field, scientists spent their days trying to imagine how a creature on a planet in the Andromeda Galaxy might appear.

One group of exobiologists, I found, held that there was a high probability that life elsewhere in the universe would look surprisingly like life here on Earth. This theory had its roots in genetic research by Cyril Ponnamperuma of the University of Maryland's Laboratory of Chemical Evolution. According to Professor Ponnamperuma, the genetic code which is responsible for determining the appearance and makeup of all living things did not, contrary to what some scientists have previously believed, originate by chance. Rather, this "key that turns the lock that opens up the door of life" (to use the phrase of one NASA biologist) evolved from a precise chemical interaction—one that would occur under the same conditions anywhere in the universe.

"There is a natural tendency, like water running downhill," Professor Ponnamperuma has stated, "for the genetic code to spell out the same words the same way every time. It's demanded by the chemistry of these compounds. . . . That means extraterrestrial life is more likely to be chemically similar to life on this planet."

Other exobiologists, however, found this to be a limited theory—as well as a dull one. "What fun are extraterrestrials," a scientist working in the NASA lab at Moffett Field complained to me, "if they turn out to be the same as us?"

NASA's Dr. Barney Oliver (an original Dolphin), for one, was convinced that evolution, whether here on Earth or another planet, would always favor functional improvements in the species. Aliens, an advanced race no doubt, would be a new-and-improved version of an old model—the human being. They would, he speculated, have bigger heads because they would need craniums large enough to house the sort of brains that would be able to process all the technological data available to them. Also, in addition to the two color-sensitive eyes in the front of their heads, he was willing to predict they would have two eyes in the back. Why? For the obvious reason. It made sense for an advanced creature to be able to see what was coming and what was going. Similarly, Dr. Oliver theorized that aliens would have more limbs than humans. And again his logic was, to a point, irrefutable: "I've always wished I had three hands."

But all the theories I heard were only so much fanciful speculation when compared to the more provocative presentation that, I learned, had occurred two months earlier in the Tank. It was a lecture based on research that had been conducted, in part, at the NASA exobiology laboratory and, not least, in Elmwood, Wisconsin.

• • •

The meeting in the Tank began, though, on a dull, unencouraging note. Colonel Phillips distributed a paper by two Canadian scientists who had worked with NASA's Moffett Field exobiology team. Its unpromising title—"Reconstruction of the Small Cretaceous Theropod *Stenonychosaurus inequalis* and a Hypothetical Dinosauroid."

Still, once one cleared away the underbrush of unfamiliar paleontological terms and awkward scientific diction, the paper had an interesting premise. The two authors wondered what would have happened if the *Stenonychosaurus,* a small, oddly graceful dinosaur with two big eyes, a good-sized brain, and highly developed three-digit raptorial claws, had not become extinct. Suppose, for the sake of argument, they said, the little dinosaurs had not vanished from Earth's surface sixty-four million years ago, but had somehow survived, and their descendants had continued to evolve. In fact, they went on, let's suppose that the descendants of *Stenonychosaurus inequalis* had achieved an encephalization quotient similar to that of Homo sapiens; that is, let's speculate (and the two scientists argued that there was no convincing reason not to) that for the last seventy-six million years or so the brains of both man and this little dinosaur had been growing at more or less the same rate. Well, the scientists asked, what would this little dinosaur have grown up to become in the twentieth century?

Interestingly enough, the paper had illustrations, and now the colonel passed these, too, around the room. The two Canadians had made a three-dimensional, sculptural reconstruction of their extinct *Stenonychosaurus*. It was a tough-looking, four-foot-high-or-so lizard with menacing claws.

Then the colonel begged everyone's patience. Before he shared any photographs of the dinosauroid, a sculpture of the hypothetical descendant of the extinct creature, he wanted first to offer a few other items for the Working Group's scrutiny.

The first was the prankish gift the CIA team had brought back from Elmwood: the sky-blue Site Center T-shirt showing a human shaking hands with an extraterrestrial.

Then, without delay, the colonel spread a half-dozen or so crude drawings over the conference table. These sketches had, he explained, appeared in *People* magazine (of all places, his frown seemed to add).

They were drawings of aliens by people from all around the country who claimed to have had close encounters.

Next, the colonel held up a book. It was Whitely Strieber's *Communion*, a best-seller about the author's experiences with alien abductors. On the cover was an artist's painting of one of the vicious humanoids Strieber had to contend with. The creature had a large, domed, hairless head, slanted, insectlike eyes, and stood about five feet high.

The aliens on the T-shirt, in the *People* sketches, and on the book cover all looked quite similar.

But, the colonel argued convincingly, perhaps that was to be expected. Rumors have a way of spreading with the force of self-fulfilling prophecies among the gullible. Often someone will "see" what he thinks he is expected to "see."

But, he challenged, how can we explain this?

He held up a black-and-white photograph identified only as "Figure 15: Model of dinosauroid head, anterolateral aspect." The colonel stated that the photograph, taken at close range, was of the head of the three-dimensional hypothetical creature created by the two Canadian scientists. It is how, he went on, two eminent scientists, men who had worked with NASA's exobiology lab, imagine intelligent life might presently look on a world where dinosaurs had the luck to escape extinction.

The creature in the photograph was nearly identical to the Strieber humanoid, the *People* alien sketches, and the illustration on the T-shirt.

It was, everyone in the Tank agreed, a most remarkable coincidence. At the very least.

And yet it was one that certainly would not have surprised any of the 991 citizens of Elmwood.

Part V
COUNTERINTELLIGENCE

THIRTY.

On a morning in the fall of 1988, two agents of the FBI's Foreign Counterintelligence division appeared at room 3E258 in the Pentagon. This was the office of the director of the Defense Intelligence Agency. And the agents had come, they announced to a startled major, because they had learned the DIA was covertly investigating UFOs.

Moments later a tense scene began to be played out. A roomful of DIA officials listened implacably to the two FBI men, yet all the time girding themselves for the inevitable second act. They would do their best to stonewall, neither confirming nor denying existence of the UFO Working Group. While the Bureau, they were certain, would do its aggressive best to solder wisps of Washington gossip into hard fact. But, as things turned out, the FBI didn't ask any questions. The two agents ended the discussion at precisely the point when the DIA was prepared to issue stiff reprimands about "need to know" and "national security issues." The FBI, it seemed, was not interested in confirming the existence of the Working Group. The Bureau had come to the Pentagon simply to ask for the group's help.

After I heard of the confrontation in room 3E258, I decided to learn why the FBI, for so long officially disinterested in UFO investigations, now appeared to be taking an active role. However, all I discovered at first was that contrary to what the incident at the Pentagon suggested, Bureau policy was still one of noninvolvement. When sightings were reported to any regional office, standard FBI procedure was to function in a liaison role, if that. Agents would gather information behind-the-scenes, and then, without making judgments, brusquely pass their reports on to more interested agencies.

It was a policy born out of pique. According to classified files (now made available under the Freedom of Information Act), J. Edgar Hoover had originally wanted the Bureau to take a large role in the flying saucer investigations, but was thwarted by an equally ambitious Air Force. When it became clear the FBI would not be the lead player, a wounded Hoover abruptly decided to walk away from the mystery. He, with a single command, established a policy of official disinterest. On October 1, 1947, after Hoover received a memo from an assistant director that the "services of the FBI were enlisted in order to relieve the numbered Air Forces of the task of tracking down all the many instances which turned out to be 'ash can covers, toilet seats, and whatnot,' " he became enraged. Bureau Bulletin 59 was issued forthwith:

"FLYING DISCS—Effective immediately, the Bureau has discontinued its investigative activities. . . . All future reports connected with flying discs should be referred to the Air Forces and no investigative action should be taken by Bureau Agents."

Still, as the stacks of voluminous files demonstrated (1,700 pages were released by the Bureau under the Freedom of Information Act), the FBI continued to receive sighting reports from all around the country. And, despite his public pose of official indifference, Hoover could not help but remain curious. In March 1950, he wrote a note to his aides ordering them to get to the bottom of things: "Just what are the facts re 'flying saucer'? A short memo as to whether or not it is true or just what Air Force, etc. think of them?"

Two months later, an Air Force memo was obediently passed on to Hoover. It made light of the phenomenon, attributing sightings to "misrepresentations" and "weather balloons." His own agents, however, continued to file less conclusive reports. One dispatch, written on July 29, 1952, noted that "the Air Force has failed to arrive at any satisfactory conclusion in its research regarding numerous reports of flying saucers. . . ." The report went on to state that according to the head of Air Intelligence "it is not entirely impossible that the objects may possibly be ships from another planet such as Mars." And still another memo, dated October 27, 1952, advised Hoover that "some Military officials are seriously considering the possibility of interplanetary ships."

More than three decades later, Hoover was long gone and his autocratic rule a receding memory, but his questions about "the facts re flying saucers" would still remain largely unanswered. And another surviving institutional relic would be the terse, almost mandarin impassivity toward UFOs that he had established as policy back in 1947. A small, though revealing example: When presidential aide Jody Powell petitioned the Bureau in 1977, the agent who handled the inquiry noted for the confidential file, "I advised him that as far as the FBI is concerned there appears to be no conceivable jurisdiction for us to conduct any inquiries upon receipt of information relating to a UFO sighting. . . ."

However, by the fall of 1988 all that was to change. The two counterintelligence agents who appeared at the DIA offices in the Pentagon had found the Bureau's "conceivable jurisdiction." It was the FBI's undisputed responsibility to investigate thefts of classified documents. And that was why, I finally learned, the two agents had come to the DIA looking for guidance. They were trying to determine whether an incredible UFO document had been stolen from the government of the United States. And, more perplexing, whether the document was genuine.

It was entitled "Briefing Document: Operation Majestic 12 Prepared for President-Elect Dwight D. Eisenhower (Eyes Only) 18 November 1952." There were eight pages. Each was stamped "Top Secret" and "Eyes Only." And it told a remarkable story. The U.S. government had recovered a crashed flying saucer and the bodies of four extraterrestrial creatures:

"A local rancher reported that one had crashed in a remote region of New Mexico located approximately 75 miles northwest of Roswell Army Air Base. On 07 July, 1947, a secret operation was begun to assure recovery of the wreckage of this object for scientific study. During the course of this operation aerial reconnaissance discovered that four small human-like beings had apparently ejected from the craft at some point before it exploded. These had fallen to earth about two miles east of the wreckage site. All four were dead and badly decomposed due to actions by predators and exposure to the elements . . . A special scientific team took charge of removing these bodies for study. The wreckage of the craft was also removed to different locations. Civilian and military witnesses in the area were debriefed and news reporters were given an effective cover story that the object had been a misguided weather research balloon."

And the story kept growing. Another saucer, nearly totally incinerated on impact, had crashed near the Texas-Mexico border. Meanwhile, on September 24, 1947, President Harry S. Truman had issued a special classified executive order authorizing Defense Secretary James Forrestal and Dr. Vannevar Bush, president of the Carnegie Institution, to begin a covert analysis of the recovered saucer and the four "Extra-terrestrial Biological Entities" (or "EBEs" to use the document's shorthand). The MJ-12 group was formed:

"OPERATION MAJESTIC-12 is a TOP SECRET Research and Development/Intelligence operation responsible directly and only to the President of the United States. Operations of the project are carried out under the control of Majestic-12 (MAJIC-12) Group. . . ."

Five years later, this group, a panel of twelve scientists, military leaders, and intelligence officials, had written the briefing paper to convince Truman's successor, President-elect Eisenhower, that the secrets they had been protecting must continue to remain inviolate.

"Implications for the National Security are of continuing importance in that motives and ultimate intentions of these visitors remain completely unknown . . . It is for these reasons, as well as the obvious international technological considerations and the ultimate need to avoid a public panic at all costs, that the Majestic 12 Group remains of the unanimous opinion that imposition of the strictest security precautions should continue without interruption into the new administration."

The briefing document also offered a tantalizing listing of attachments. Attachment A was a three-sentence memorandum from President Truman to Secretary of Defense James Forrestal emphasizing that "considerations relative to this matter should rest solely with the Office of the President . . ." The remaining seven attachments ("Operation Majestic-12 Blue Team Report #5; Operation Majestic-12 Contingency Plan MJ-19949-04P/78," among others) were missing.

But that caused little concern. The eight pages were startling enough.

The Working Group held a crash meeting to discuss the documents. Yet, oddly, a long while passed at that late-night session—"years, an eternity," it would be recalled with some amazement—before anyone had the nerve to confront head-on the obvious question. Instead, the group, cautious professionals in all matters, danced carefully around its edges.

There was a great deal of probing discussion, for example, about the form and placement of the classification stamps, the pagination style at the top of each page, and even the single-spaced typing of the report. Everyone, it seemed, had an opinion. At last, after much recondite analysis, it was grudgingly agreed that, in terms of style and form, the pages at the very least *appeared* authentic.

And then there were no longer any diversions left to pursue. It was time to take the leap. The only truly important question was asked: "Are these documents real?"

"That's precisely," Colonel Phillips answered, "what the FBI wants to know."

And now, so did I.

THIRTY-ONE.

Many of those familiar with the FBI's investigation into the authenticity of what came to be known as the MJ-12 documents insisted the case began in June 1987. After all, they pointed out with solid reason, that was when the letter demanding an official inquiry was written to the Bureau.

But Phil Klass, as exacting and as quarrelsome as always, argued that it would be more accurate to take a more historical approach. He suggested to me that if certain events had not occurred twenty-one years earlier, in August 1966, then most certainly the letter would never have been written. And, though he was too polite to point this out, his opinion had to be given some consideration. It was he who wrote that fateful letter to the FBI in the first place.

When Phil Klass awoke on the hot, muggy morning of August 22, 1966, he had little idea of what was in store for him. Yet by the day's end, his life would be irrevocably changed.

It was not a transformation he would ever have wished. Klass was very comfortable, content even, with the way things had worked out. At forty-six, a few steps into the valley of middle age, he had finally managed, he felt, to steer his career into an arena that was exciting and intellectually fulfilling. He got paid to learn about new things every day. What could be better than that? The grim and unrewarding decade he had put in toiling for General Electric, the only job available after he had graduated with his electrical engineering degree from the University of Iowa, was now only an unhappy memory. He had made his escape fourteen years ago after reading a help-wanted ad in a magazine. The editors of the Washington-based *Aviation Week* were looking for a technical writer and

Klass, without much hope but certainly game, sent in his résumé. He got the job, got out of G.E., and got out of the Midwest. As suddenly as if he had been granted a wish, life became joyous. His beat was the aerospace industry, and in a town caught up in the space race boom, it was fertile turf. Better, he was a natural. He brought to each assignment a reporter's impudent curiosity, an editor's uninhibited cantankerousness, and the sort of relentless energy that is a product of that rarest of commodities in journalism—genuine interest. Another plus (and one never to be underestimated): he was a bachelor, and without the confining discipline of a domestic life he could allow the demands of his work to set his pace. Of course, he was a success.

And then, just as his future seemed to be moving cheerfully and inevitably along, a series of small, unforeseen events occurred in the midst of that suffocatingly humid summer of 1966. The first was his happening—a casual, almost accidental act at best—to begin to skim through a copy of John G. Fuller's *Incident at Exeter*. The book was an account of Fuller's affirming investigation into the flurry of UFO sightings that had occurred in the prep-school town of Exeter, New Hampshire. To this day, even from the clarifying perspective of hindsight, Klass cannot fathom why he first picked up the book. He had no interest in UFOs; he was a reporter and, like most in his insider's world, a cynic. Even for recreation, his inclination was to give his mind a workout with heavyweight science, to pick up texts thick with a ton of facts and data. He, always busy, never idle, had no time for whimsical speculations about flying saucers. Yet Klass, undeniably, began browsing through the book. In a matter of pages, he was hooked. Soon he was reading very carefully. And the more he read, as he furiously jotted notes in the margins of page after page, the more his indignation grew.

Fuller was wrong! The clue was the power lines. Klass, the armchair detective, deduced that while Fuller's eyewitnesses were convinced they had observed spaceships, their close encounters had actually been with something a lot more mundane—ball lightning. Since all but one of the sightings occurred near high-tension power lines, Klass realized that what the people in Exeter had seen was flashes of radar energy being generated by those overhead cables. Case solved, he told himself triumphantly.

But Klass from the first also knew he could not let it go at that.

It wasn't so much that Fuller had to be publicly humiliated; any man, Klass readily conceded, was entitled to his mistakes. Nevertheless, it was a matter of honor. Klass felt he had an obligation to set the record straight. In the realm of ideas, he lectured himself, it was the duty of every knight to slay false dragons. So—another first—he decided to write a book review. Initially, his thought was to see if *The Washington Post* was interested. But when he mentioned this plan to his editor, Klass was told, "Forget them. You work for us."

Phil Klass's review of *The Incident at Exeter* appeared in *Aviation Week* on August 22, 1966. Its publication made him an instant authority on the phenomenon of unidentified flying objects. He did not know what he was getting into.

"Don't do it," warned Robert Hotz. "You don't want to get involved with all those kooks." That was his editor's reaction when his star reporter told him he was contemplating doing another article on UFOs. But Klass was beyond taking anyone's—even his boss's—advice. His was a just and perhaps even holy war. There were intellectual wrongs to be righted.

The subject (or, more accurately, target) of this second article for *Aviation Week* was a thick volume of five hundred confirmed sightings as documented by the National Investigations Committee on Aerial Phenomena (NICAP). It was a group of true believers, and Klass took after them with a devilish and unrestrained joy. Once more, science was his lance, and he used it to skewer page after page of undisciplined assumptions.

Regardless of whether or not Klass's determined assault bruised anyone's faith, the carefully reasoned article certainly did attract attention. All at once, it seemed, Klass was showered with requests from other magazines to do articles on UFOs, to appear on radio shows, even television. Eager to proselytize heathen minds, he gladly accepted more than a few of these offers. He became, at least in this small yet vociferous world, a celebrity. There was no turning back; it was impossible to abandon his crusade because his sudden silence would have been viewed as capitulation. For by now, as if overnight, he had been christened by both friends and foes "the country's leading UFO debunker."

And the invitation to Thanksgiving dinner seemed only to confirm his anointment.

It was a story Klass had told many times over the years, and yet even after two decades as a staple of his repertoire he was still enough of an Iowa hayseed to deliver it with a sincere, imagine-that gloss.

"I received a telephone call," Klass has written about the incident that occurred in 1966, just three short months after his first review had appeared in *Aviation Week,* "from Random House's founder and chief executive, Bennett Cerf . . . Cerf invited me to join him and 'a few friends'—including Frank Sinatra and 'Mike' Cowles, whose publishing empire included *Look* magazine—for Thanksgiving dinner at his New York townhouse, and I accepted. Following dinner, Cerf assembled about twenty of the male guests and asked me to speak to them on UFOs. One of these was Cowles, who, I later learned, was favorably inclined toward UFOs and had often tried to convince Cerf. After my talk, Cerf invited Cowles to challenge my skeptical views, but he declined. Not surprisingly, Random House published my first UFO book in 1968. . . ."

It did not, of course, end with that one book, *UFOs: The Public Deceived.* For the next twenty years, Phil Klass balanced the demands of his hectic and vigorous career at *Aviation Week* with his private and consuming need to demystify UFOs. It was a nearly around-the-clock schedule. He, by now a senior editor, would work eight hours at his desk at the magazine; and then he would hurry home to his desk in his downtown Washington town house where, fueled by black coffee, he put in another six or even eight hours trying to prove that what people think they have seen is quite often not at all related to what they actually have witnessed. It was detective work, a unique sort of problem-solving, and his curious mind enjoyed the challenge. But it also was a responsibility and, at times, a burden.

There was, to cite just one incident that nearly broke him, the time in 1986 when he, an energetic sixty-year-old, was working full-tilt on a special report on electronic countermeasures for the magazine. There was so much to learn before he could even begin to write that Klass found himself working on Christmas Eve; and, more grief, he wound up spending New Year's Eve at his desk, too. He was logging, he remembered, intense sixteen-hour days to prepare that report. And then in the first week of January, just when his deadline was looming menacingly, the phone calls began. First AP, then UPI, then the TV stations, plus at least a dozen or more news organizations from around the world. The pilot of a Japan Airlines 747 cargo plane had reported an encounter with a giant UFO over Alaska; what did the world's leading skeptic think of *that*?

Klass's immediate reaction was to dismiss all inquiries with a polite "I'm sorry, but I'm too busy to look into that case right now." But if he did that, he quickly realized, UFO believers all around the world would be exulting, "Aha! We finally got Klass to throw in the towel." He had no choice but to go to a twenty-hour-a-day schedule.

It was worth it, for, after laboriously reviewing Federal Aviation Administration reports on the sighting, detective Klass was convinced he had solved another case. On the night in question there was a nearly full moon approximately twelve degrees above the horizon, and the bright moonlight reflecting off turbulent clouds of ice crystals must have generated the "undulating flame-colored lights" the pilot had seen.

But this was only one case, of thousands. There was so much to do, and he was just one man—and growing older. For a while, in the 1970s, it had seemed as if he had recruited two young disciples, but this pair of scientists, after a few years of intense apprenticeship, had abandoned ship. "Who could blame them?" Klass conceded. "It's demanding, grueling work, and there are no financial rewards. In fact, all you get is abuse from the kooks who think you're working for the CIA."

So Klass, by nature a loner, convinced of the justness of his cause, continued on his own. Even when he finally gave in a bit and got married at the age of sixty, he made a point of telling his bride, six years his junior, that she would have to be tolerant of his crusade. There still were many more battles to fight.

There Klass would be, then, at UFO conferences around the country; "the skunk at the garden party," he would boast with a relishing laugh. Believers would parade by wearing T-shirts proclaiming "Klass is an Ass" and he would love it. He did not hide. He got a kick out of being a well-known enemy. A jaunty, elegant, pencil-thin figure with a scholar's thick black horn-rims and a rake's straw boater, he would appear at conference after conference, his pocket-size tape recorder aimed like a weapon at the often flustered speakers. "I enjoy playing the villain," he would admit. "The louder they bellow, the better a job I guess I'm doing."

In his long career as the "country's leading debunker" Klass had tackled all sorts of cases. Time after time, he had found nothing more mysterious than the planet Jupiter in the night sky where others had seen flying saucers. When in the early 1980s another troop of believers began insisting that "missing time" was the first indication that a person might have experienced a covert UFO abduction, he found a more prosaic explanation: "Is there anyone who has not at some time looked at a clock or watch and discovered that it was much later than expected . . . thus experiencing 'missing time'?" He had, he was confident, solved beyond a reasonable doubt every case that had come his way.

Yet just as he was approaching his seventieth birthday, just as he was for the first time beginning to think seriously about slowing down, came a new challenge. It might very well be, he realized with a small pang, his last case. His final chance to pull back the curtain and reveal—an empty stage. And his final chance to play the villain.

But, he rejoiced, it was a worthy finale. For as soon as he read the MJ-12 documents he understood what was at stake. Either he would harpoon at last the great white whale, or he—and his cause—would be done in. "Either," as he wrote, the documents were "the biggest news story of the past two millennia or one of the biggest cons ever attempted against the public . . ."

And so, carefully, ploddingly, drawing on a lifetime of experience, Klass, at the peak of his talents, began his investigation into the MJ-12 documents. It was while still caught up in this false calm, the sense of controlled emergency that accompanied the onset of each new case, that Phil Klass took the time to write a letter. It was dated June 4, 1987, and it was addressed to Mr. William Baker, Assistant Director,

Office of Congressional and Public Affairs, the Federal Bureau of Investigation:

Dear Mr. Baker:

I am enclosing what purport to be "Top Secret/Eyes Only" documents, which have not *been properly declassified, now being circulated . . . by William L. Moore . . . Burbank, Calif. 91505 . . .*

THIRTY-TWO.

Meanwhile across the country in Los Angeles, Bill Moore was busy conducting his own investigation of the MJ-12 documents. Like his adversary Phil Klass, Moore understood that at last, after so many small, inconsequential skirmishes, the Great Battle was at hand. At its end, either the government's cover-up would be finally exposed and the reality of UFOs established, or all—Ufology *and* his reputation—would be in ruins. Proving the authenticity of the MJ-12 papers was, he would insist in an outburst both rare for its candor as well as its volume, his "personal search for the Holy Grail." Like Klass, he was totally committed.

It was, then, a grudge match of ambitions; and the two men were, at first glance, obvious opponents. Klass, the tweedy and professorial senior citizen, had a well-honed air of bourgeois normalcy surrounding him, of a long and steady life spent acquiring such rewards as a pension, massive insurance, and a thirty-year mortgage. With his tie firmly knotted at his neck, a gentleman's trilby on his head, and the handle of his briefcase resolutely gripped, he was another one of those busy men who help make Washington such a busy city.

Moore was a more freewheeling sort. He took life one day at a time, and after forty-five years it had sometimes proved to be a pretty bumpy trip. He had gone through a divorce, a bankruptcy, and just last month, once again down to pennies, he had had to let his Blue Cross lapse. He had managed, though, to hold on to his ten-year-old Harley and that was worth all the tangled economies the scheme had required. The bike was in many ways an essential part of him; it made him complete. A compact man with a barrel chest and forearms as stout as beer kegs bulging from his T-shirt, a Michelob key ring dangling from his jeans, a black leather jacket and heavy, stomping boots, he looked at home and, oddly, at peace while mounted on his revving Harley. Even off his motorcycle, Moore cultivated a biker's sullen, don't-mess-with-me demeanor. With one big hand pensively stroking his blondish beard, he would narrow his eyes, flare his broad nostrils, and squint across the table at a questioner as though sighting him in the scope of a gun. And when he spoke, his voice would sound as if it had traveled from the depths of his stomping boots, a faint and labored monotone, each slow syllable apparently offered only as a small concession. He was—and he was not squeamish about making this clear—a man with an attitude.

Yet, despite such differences in their armor, Moore and Klass brought into battle many of the same resources. Both men were inventive and exhaustive researchers, so self-possessed that they would readily devote decades to establishing a single, arcane fact if it would help anchor their grand and conflicting theories. And both men shared the same weakness—their absolutism. They were convinced of the inherent rightness of their opposing causes, and they were completely dedicated to one another's downfall.

So when Bill Moore learned about Phil Klass's efforts to expose the MJ-12 documents as a hoax, there was not even a flutter in his resolve. "Let Klass rant all he wants," he announced. "It is the achiever who is remembered and honored by history," he reminded his supporters, "and not the debunker." Similarly, when Moore learned about the FBI investigation, that agents from the counterintelligence squads in both the New York and Los Angeles bureaus were asking questions about the documents and about him, he was also undisturbed. The federal inquiry into his activities was nothing short of ironic. It wasn't that long ago that government counterintelligence agents had recruited him to spy on his friends. And, further cause for a small sardonic smile, he was certain that if he hadn't taken part in that covert operation, the MJ-12 papers would never have come his way. In fact, it was more than likely the top-secret papers would simply never have surfaced.

It was 1979, and it was a great year, possibly the greatest time of Bill Moore's life. Success had been truly emancipating. On the strength of his having cowritten *The Philadelphia Experiment*, a best-selling novel that told a shrewdly provocative story about a voyage to a futuristic world, Moore made the decision (and had the funds) to embark on his own liberating journey. He packed up his wife and four children, waved a perfunctory good-bye to icy Minnesota and the slavishly regimented decade he had put in as a high-school English teacher, and headed off to the sun and free-lance freedom of Arizona. He would devote all his time to writing. But there was more: He would unite, as term after term Robert Frost had advised his students, his avocation and his vocation. He would write full-time about what had previously been a cherished but only part-time concern—UFOs.

Moore had become interested in the phenomenon because of a hoax. As a teenager growing up in Pittsburgh, his ambition was to be an astronomer and dutifully he attended night classes at the Bule Institute of Popular Science as well as the monthly meetings of the Amateur Astronomers Association. It was at one of these meetings that a bold guest speaker devoted the hour to a discussion of flying saucers. For young Moore, the possibility that people from other worlds were visiting Earth was not just fascinating, it was compelling. He had to learn more. The first book he read was Frank Scully's celebrated *Behind the Flying Saucers*. It was a real page-turner, full of "true" stories about crashed flying discs in New Mexico and recovered humanoid bodies. Unfortunately, as an older and wiser Moore would readily volunteer, it wasn't true. Still, at an impressionable time in his life the book had performed a valuable intellectual service. Moore now suspected something was out there.

He got more substantial proof when, after graduating from Thiel College in Greenville, Pennsylvania, he took a teaching job in upstate New York. In college, he had joined NICAP, and now the group called on Moore to investigate a flurry of sightings in Garrison, New York. It was 1966, just about the time when Phil Klass was busy debunking a thick bookful of NICAP's eyewitness reports, and, as it happened, so was Bill Moore. All the close encounters he reviewed, Moore found, were easily explainable and easily dismissed. Except for one. There was one case that, no matter how he twisted the evidence, grilled the witness, or studied the astronomical data, still came out unsolved. He couldn't say for certain that there had been *something* darting about the night sky above Garrison. But, just as firmly, he wasn't sure there had been *nothing*. And that, too, was a source of wonder for many years.

Therefore, when in 1979 he set off to devote all his energies to investigating UFOs, it wasn't so much that his mind was already made up as that it was leaning in a very definite direction. But after he threw himself into researching the government's cover-up of the strange events that had occurred in Roswell, New Mexico, in 1947, all his lingering doubts were overwhelmed.

The Roswell Incident, which Moore again coauthored with Charles Berlitz, was published in 1980. It was an abundantly confident exposé, and yet at the same time a total muddle. Bizarre allegations, fanciful hypotheses, and documented facts were cut and pasted together as though sharing the same pedigree. Even Moore himself was not pleased with the final product. He broke with Berlitz (who went off to write singlehandedly a best-seller about the Bermuda Triangle) and instead teamed up with Stanton Friedman, a nuclear physicist and, equally valuable qualifications for the battles that lay ahead, a nearly compulsive ferreter of the obscure as well as a staunch believer in the reality of UFOs. Together, they would spend almost another decade studying the events at Roswell. They would interview ninety-two people, a list that included, as Friedman would be quick to point out, "thirty-one who had first-hand involvement." And now, when the tabloid embellishments and leaps of reason that had marred the book were scraped away, a story could be meshed together that at the very least was perplexing.

There was no doubt that on July 2, 1947, a rancher, William Brazel, discovered widely scattered wreckage about seventy-five miles northeast of Roswell, New Mexico. Brazel contacted the local sheriff. Two days later Major Jesse Marcel, a staff intelligence officer at Roswell Army Air Field, was sent to investigate. Years later, after having retired as a lieutenant colonel, Marcel would recall what he had found:

"There was all kinds of stuff—small beams about three-eighths or a half-inch square with some sort of hieroglyphics on them that nobody could decipher. They looked something like balsa wood, and were of about the same weight, except they were not wood at all. They were very hard, although flexible, and would not burn."

The major's son also saw the wreckage. Thirty-four years later, now a flight surgeon for the Air National Guard, his memories, too, remained vivid:

". . . what I saw in 1947 is unlike any of the current aircraft ruinage I have studied. This craft was not conventional in any sense of the word, in that the remnants were most likely what was then known as a flying saucer that apparently had been stressed beyond its design capabilities. I'm basing this on the fact that many of the remnants, including the eye-beam pieces that were present, had strange hieroglyphic type writing symbols across the inner surface. It appeared to me at the time that the symbols were not derived from the Greek or the Russian alphabet, nor of Egyptian origin with their animal symbols."

And it was uncontested that when Major Marcel returned to the base with the wreckage an official press statement was released:

"The many rumors regarding the flying discs became a reality yesterday when the intelligence office of the 509th Bomb Group of the Eighth Air Force, Roswell Army Air Field, was fortunate enough to gain possession of a disc through the co-operation of one of the local ranchers . . .

"Action was immediately taken and the disc was picked up at the rancher's home. It was inspected at the Roswell Army Field and subsequently loaned by Major Marcel to higher headquarters."

Further, there was no dispute about the text of the official (that is, *public*) reaction from these "higher headquarters." Brigadier General Rodger M. Ramsey, commanding the Eighth Air Force at Fort Worth, Texas, announced in a radio broadcast that the wreckage was simply the remains of a high-altitude weather observation device. "There is nothing to get excited about," he said.

Bill Moore and Stanton Friedman, however, remained unconvinced. Forty years would pass and they were still, despite the general's appeal for calm, very excited. "Ten first-hand witnesses are on record as having stated that the object was . . . some sort of space craft," Friedman would write in 1987. "Five persons who actually handled pieces of the wreckage have described very unusual symbols on pieces of the wreckage."

The government, both Moore and Friedman believed after years of doggedly tracking down witnesses, was hiding the truth—an alien craft had crashed near Roswell, New Mexico.

It was largely the position that, a bit more shrilly and with many more of the pieces in the puzzle still missing, Moore had first taken in *The Roswell Incident.* The book, however flawed, however much an awkward first step, sold quite well. It made Moore a figure of some renown in the UFO community, which was something he had been hoping for as he set out on his new, free-lance life in Arizona. But there was another unanticipated consequence, one that was beyond his wildest dreams, and was, at first, even more intimidating than his most secret fears. The book persuaded government counterintelligence agents to recruit him as a spy.

The first contact was made in Washington, D.C. It was September 1980, and Moore had just finished a radio call-in show on WOW. He had talked at length about the Roswell crash cover-up, and he was quite pleased at the way it had gone. He had been, he felt, very persuasive.

It was in this self-congratulatory mood, while walking briskly through the radio station's lobby, his mind looking forward to an afternoon of digging for more clues at the National Archives with Stanton Friedman, that the security guard called out to him.

"You Moore? There's a phone call for you."

Moore took the receiver. The man on the other end introduced himself as a colonel, and Moore paid little attention to his name.

The colonel went on matter-of-factly: "You're the only person we've ever heard who seems to be on the right track," he said.

Moore wondered who the "we" could be, but instead of asking he simply thanked the colonel for phoning and said good-bye. He was eager to meet Friedman and get down to some serious work.

He did not give the call another thought until about a week later. He had just completed an appearance on KOB's morning show in Albuquerque, when, again just moments after he was off the air, he was informed there was a phone call.

Moore, without a trace of suspicion, said hello. When the caller spoke, however, he quickly realized that something very strange was happening. Moore did not recognize the voice, but the words were an echo: "You're the only person we've ever heard who seems to be on the right track."

Moore's mind was racing. He was now 1,500 miles from Washington, halfway across the entire country, the smooth voice on the line was a new one, yet the words were identical. *What was going on?*

When the caller casually suggested they meet that morning in a coffeehouse in downtown Albuquerque, Moore did not argue.

"You can't miss me," the caller said before hanging up. "I'll be wearing a red tie."

It was on his way over to the coffeehouse, dozens of possible scenarios filling his imagination in the course of the brief yet feverish drive, that Moore came up with a plan of his own. With an instinctive cunning, he decided to change the rules for this meet. Rather than wait obediently at a table, he felt it would be wiser if he controlled the situation. His plan was to park his car across the street from the coffeehouse and watch who entered. That way, he would be able to size up the man with the red tie before proceeding. At the very least, he hoped, such caution might provide him with a small advantage.

It was a good plan; it just didn't work. Moore was sitting in his car, staring out at the coffeehouse across the street, when there was a hard tap on the window on the passenger's side. The noise was so unexpected, it made him jump. Recovering as best he could, he turned toward the sound. Standing there was a man wearing a red tie.

"You Mr. Moore?" he asked, but his sly smile said he didn't have the slightest doubt.

"How did you know?"

"It was easy," said the man as though that was explanation enough. Then he opened the door and, without waiting for an invitation, climbed into the front seat.

That was how Bill Moore first met the man he would come to call the Falcon.

In the weeks that followed that initial encounter, the Falcon, offering a sliver of information one day, a little more the next, wove a story that left Moore excited, and a bit stunned. The Falcon, according to a determinedly hush-hush and protective Moore, "was a well-placed individual within the intelligence community who claimed to be directly connected to a high-level government project dealing with UFOs." This group, too, was uncomfortable with the government's continuing cover-up. Its members also wanted to get the full story about UFOs out to the public. And so it came about, after many months of tentative probing, that the Falcon offered Moore a deal. The group would help Moore with his research, if Moore would cooperate with them.

"Cooperate?" Moore asked.

The Falcon said all the details would be explained by someone else. But first Moore would have to volunteer to help.

Moore, intrigued, on guard, yet coolly pragmatic, offered his services to his country. Soon he met the man he would identify as his liaison to the Falcon—Master Sergeant Richard C. Doty, an agent serving with the Air Force Office of Special Investigations (AFOSI). And it was Doty, then assigned to Kirtland Air Force Base in Albuquerque, New Mexico, who in 1983, Moore claimed, gave the new recruit his first undercover assignment. Moore was to infiltrate Project Beta.

THIRTY-THREE.

With the sun beginning to set, the humpback Manzano foothills would cast long, broad shadows across Coyote Canyon. The sky would slowly start to bleed, turning from a deep, brilliant desert blue to a pastel shade, a faded denim color streaked with an irradiating red, until, at last, it all settled easily into a soft zinc gray. And then the lights would appear. In these last moments before the New Mexico night began, coming from somewhere in the west near Kirtland Air Force Base, the strange craft, their running lights aglow, began their maneuvers. They would fly in a circling formation in the dusk sky above the Manzano Nuclear Weapons Storage Facility, and next fly south toward the Coyote Canyon test area. Every evening they came. Their arrival was as regular as the sunsets, and no less spectacular.

From the deck of his house perched high in the Four Hills section of Albuquerque, Paul Bennewitz had a perfect view. Night after night, he paced the deck, an eight-millimeter movie camera in his hand, as he, with considerable anxiety, recorded the erratic, hovering flight paths of these craft. At the same time, his tracking antennas would also be at work, sweeping in unison across the sky. With lumbering deliberativeness, the huge antennas automatically rotated toward the ships, vectoring in on their flight. They moved clockwise, their rotors loudly grinding, until contact was made. Then banks of ultrasensitive receivers—lovingly handcrafted machines, the cherished brainchildren of Bennewitz's own ingenious designs—would come alive. A steady, low-frequency electromagnetic *beep . . . beep . . . beep* would fill his workroom. The signal came in modulated pulses, loud and clear and well-defined like the exultant opening chords of Beethoven's Ninth Symphony. He never doubted those strange craft were sending a message.

Each night it was all recorded. There were over 2,600 feet of film. A locked filing cabinet held the tapes of months of encounters. These were Paul Bennewitz's clues and, after much painstaking analysis, his proof. They were the irrefutable cornerstones of his great discovery—Project Beta.

The insights that culminated in Project Beta, Bennewitz's grand theory about UFOs, first surfaced under hypnotic regression. It was 1979 and Dr. Leo Sprinkle, a New Mexico psychologist, was challenged by the story a deeply troubled female patient had revealed under hypnosis. From the very depths of her subconscious, she had purged herself of a most remarkable tale. She had been abducted by aliens.

Dr. Sprinkle believed her, and he did not believe her. Filled, then, with a sense of concern and fascination, as well as skepticism, he decided to consult his friend Dr. Paul Bennewitz. It was his hope that Bennewitz, an accomplished physicist, a prodigious inventor, a man of science with a wizard's mind as well as a soft, sympathetic spot for all talk about UFOs, might be able to contribute some insightful analysis. Bennewitz was most definitely interested, and the woman was eager for whatever help he—or anyone—might offer. It was arranged that Bennewitz would be present during her hypnosis.

The sessions continued for three months. The woman would slip into a trance easily, her eyes nearly closed, her voice a low monotone; and then, under Dr. Sprinkle's prodding, more and more of her repressed encounter would come forth. Moments before her abduction, she recalled as the two men listened, she had witnessed a bizarre ritual. The aliens were killing cattle, draining the beasts of their blood. She saw it all. That was why they took her. They took her to their ship and she was forced to watch as the aliens did strange things, things she couldn't quite understand, things she still didn't quite recall, with the cattle and with the blood. And then they did things to her.

As Paul Bennewitz listened over the course of those months to the woman's agonized tale, he did not at first know what to make of it. But the more he mulled it over, he became convinced—absolutely certain beyond a shadow of a doubt—that she was telling the truth. No one, he felt, could be that good an actress. Her pain was genuine.

But there were still crucial pieces missing from her story. It was necessary, Bennewitz realized, his own fears building, to learn what the aliens had done to this poor woman. He urged Dr. Sprinkle on. The facts, however, were buried too deep, were too successfully repressed. Yet Bennewitz was unyielding. He was convinced those lost moments aboard the spacecraft were the keys to understanding the motives of the aliens. His task was apparent. What the victim couldn't remember, the rescuer—and by now he saw himself in that role—would discover. So piece by piece, part observation, part scientist's logic, part instinct, he over many months came to an understanding about what had happened. The aliens had surgically implanted mind-control devices in the woman's skull. They could see what she saw. They could hear what she heard. They could control her every move.

Bennewitz was terrified.

Still, goaded on by what was at stake, in a state of constant alert, he conceived Project Beta. His careful and documented monitoring of the alien ships flying over the New Mexico desert, and the messages they were sending to control their victims, began.

From the start, rumors full of mystery and promise involving Project Beta swirled through the tightly bonded communities of kindred thinkers who lived across the Southwest. And so, looking at subsequent events from this perspective, perhaps it was inevitable that Bill Moore's and Paul Bennewitz's paths should cross. Nevertheless, it wasn't until 1981, after Project Beta had been in operation for nearly two years, that a curious Moore, now a director at the Aerial Phenomena Research Organization (APRO), a Tucson-based group of UFO investigators, drove east from Arizona to the scientist's home in Albuquerque. His assignment was to evaluate Bennewitz's findings.

Moore, who prided himself on his ability to size people up, found much to admire in the scientist. Bennewitz, mesmerizingly articulate, able to pepper any conversation with seemingly inexhaustible flourishes of esoteric information, had the confident manner of a man who had grown up being the smartest boy in the class. And yet there was also something disquieting about him. Moore found his intensity—a trait many would agree Moore could analyze with considerable authority—especially disconcerting. It was as if Bennewitz felt his role was to serve as one of history's not so silent witnesses; or, perhaps he even saw himself as a prophet, one of those high-minded souls whose nagging earnestness was meant to call lesser lives into question. Whatever it was, it rankled Moore. He preferred to take Bennewitz in small doses.

As for Project Beta, Moore viewed the footage of the hovering lights and listened to the tapes of recorded messages. It was undeniable that Bennewitz had seen and heard something; the films clearly depicted unusual lights maneuvering near the Sandia National Labs complex, a classified Department of Energy facility on the Kirtland base. And, just as certainly, Bennewitz's receivers had been monitoring odd low-frequency electronic signals. But Moore was not at all convinced that these "discoveries" had anything at all to do with UFOs. The strange craft might be, he reasoned, nothing more ominous than Air Force helicopters or perhaps even some sort of experimental plane. Similarly, Moore found it difficult to accept that the signals were alien radio transmissions. Bennewitz's highly touted computer-generated decoding program was based, as best Moore could tell, on the sort of shaky assumptions that would just as readily have translated the pulses of Morse code into an extraterrestrial monologue.

Moore returned to Arizona and announced to APRO that as far as he was concerned Bennewitz was a dedicated researcher who just didn't seem to have the emotional objectivity to sort, as he noted with deadpan candor, "the shit from the candy." Still, over the years Moore remained in touch with Bennewitz and the two men became friends; after all, they were involved in the same quest. And it was with a mixture of amusement and bewilderment that Moore watched as Project Beta evolved into an all-encompassing theory. What had started with some fragile conjectures about mind-controlling aliens had, Moore would state with a sigh, "blossomed into a tale which rivaled the wildest science fiction scenario anyone could possibly imagine."

According to Bennewitz—and supported, he insisted with unshakable ferocity, by his research—two opposing forces of aliens had invaded the United States. The white aliens wanted intergalactic brotherhood; they came to this planet in peace. However, the malevolent group, the grays, were in control. It was the grays who were responsible for the cattle mutilations, the human abductions, and the implanting of mind-control devices in humans. The government was not only aware of this, but had also negotiated a secret treaty with these invaders. The grays were granted the right to establish an underground base beneath Archuleta Peak near Dulce in northwestern

New Mexico, and in return the military had received a shipment of extraterrestrial weapons. But then an atomic-powered alien spaceship crashed on Archuleta Peak. The grays suspected sabotage. And, Bennewitz was convinced after decoding radio transmissions, the treaty was about to be broken. The angry grays were preparing for nothing short of total war.

It was a theory that Bennewitz, in his own mind another Paul Revere, was devoted to circulating. He attempted to contact not just UFO researchers like Moore, but also congressmen, military commanders, members of the scientific establishment, and even the President. "Instead of withholding judgment until all of the facts were in, Paul insisted on repeatedly going off half-cocked to anyone who would listen," Moore complained. The way Moore saw it, Bennewitz was "his own worst enemy."

It would not be until months later, after Moore was recruited by the Falcon and given his assignment by Air Force Office of Special Investigations agent Richard Doty, that Moore would realize Bennewitz had a more formidable enemy—the government of the United States.

Disinformation, as the Soviet term *desinformatsiya* was quickly anglicized by admiring Western intelligence agencies, is the propagation of false, incomplete, or misleading information to targeted individuals. But for a disinformation campaign to be truly successful, it must accomplish two related goals. One, the target must act on these new "facts." And two, the target must be irrevocably diverted from the more fruitful path he had previously been following.

For the past three years, since 1980, Bill Moore learned from AFOSI agent Doty, counterintelligence officers from a variety of agencies had been running a disinformation campaign against Paul Bennewitz. The purpose of the exercise—or so Moore would remember being told by a gloating Doty—was systematically to confuse, discourage, and discredit Bennewitz.

Their work had been remarkably successful. It was government agents, pretending to be friendly coconspirators or using other, more convoluted covers, who had first passed on to a gullible Bennewitz "official" documents and stories detailing the secret treaty between the U.S. government and evil aliens, the existence of underground alien bases, the exchanges of technology, the wave of brain implants, and even the tale about the spaceship that had crashed into Archuleta Peak. These "facts" became the linchpins of his grand theory; and, fulfilling all the government's hopes, Project Beta—the filming of airship maneuvers in the vicinity of nuclear bases and the monitoring of the unusual signals emanating from these craft—had been now relegated to a secondary concern.

And now agent Doty wanted Moore to join the government's team. He assigned Moore to spy on Bennewitz. Moore's job was to report on a regular basis to Doty about the effectiveness of the government's disinformation campaign. Did Bennewitz still believe all the wild tales that had been passed on to him?

For four years Moore kept a careful watch on his friend. For four years he listened mutely as Bennewitz complained that his phone was tapped, that his office had been broken into. Moore, the dutiful recruit, even passed on to Bennewitz the "Aquarius Document," an actual classified AFOSI message that had been skillfully doctored—by Doty? the Falcon? Moore never asked—to prove that an alien invasion was at hand.

And for four years Bennewitz never suspected his friend of any sin worse than skepticism. Moore was the perfect spy.

Why? That was the one question Moore kept asking himself as he, now an insider, observed the government's sustained campaign against Bennewitz. Why were Doty and the Falcon so intent on discrediting one solitary UFO crusader?

The truth was never explained to Moore. He wondered if AFOSI had simply picked Bennewitz at random, that he was an unlucky target of an ongoing counterintelligence teaching exercise. Or, perhaps Bennewitz had actually been filming UFOs from his sun deck; the government's long cover-up was jeopardized and, therefore, Bennewitz—and his film and tapes—must be discredited at all costs. Or, equally plausible, it was possible that Project Beta had been monitoring a top-secret military training program, and a plan to discourage anyone else—foreign spies as well as believers in UFOs—from paying too much attention to these maneuvers was quickly conceived. Moore would never know.

But whatever the reasons behind it, Moore had no doubts about the effectiveness of the government's disinformation program. Moore watched as Bennewitz was driven to the breaking point. As he was fed stories about evil and threatening grays, Bennewitz grew more emotional. He kept guns and knives hidden throughout his house. He had extra locks installed on his doors. He could not sleep. He turned his business over to his son. At lunch with Moore, Bennewitz, his hands shaking, his face as haggard as a skeleton's, told his friend that aliens were coming through his walls at night and injecting him with hideous chemicals. The chemicals knocked him out; he was very worried about what the aliens had done to him when he was unconscious. As he spoke, he smoked constantly. Moore, whose job was to be observant, counted each of the twenty-eight cigarettes Bennewitz had puffed in the course of the forty-five-minute meal. It was not long after that lunch that Bennewitz was hospitalized for exhaustion and fatigue.

And Moore, how did he feel about his tacit complicity in the government's plot against Bennewitz? Did he feel ashamed by his silence? By his betrayal of his friend?

He has yet to comment, and his reluctance is understandable. Instead, he preferred to describe his work with Doty as an "opportunity," his spying on Bennewitz as "the price I had to pay." And, if one looked at it in such hard, pragmatic terms, it was a moment of high achievement. Moore had penetrated a cadre of top-level U.S. intelligence agents who were involved with UFOs. His course was set: "I would play the disinformation game, get my hands dirty just often enough to lead those directing the process into believing that I was doing exactly what they wanted me to do, and all the while continue to burrow my way into the matrix so as to learn as much as possible about who was directing it and why."

THIRTY-FOUR.

It was a long and complicated courtship. The Falcon, Doty, and their associates were very cautious, very demanding. But never was there a more obliging, a more eager-to-please suitor. Bill Moore was on the inside. He felt he had been handpicked by a group of intelligence operatives working, or so they not very subtly kept implying, on a high-level project involving UFOs. Let them toy with him. Dangle clues flirtatiously. Send him on wild-goose chases. He would simply smile, and trot back the next day for more. He was prepared to do anything to win their hearts, their trust—and their secrets.

Not that he didn't have his own secrets. Everyone involved in this elaborate counterintelligence game, he decided, was acting out some predetermined role, so Moore, a born field man, had adopted his own cover. He would be, he later said without a trace of irony, "the useful idiot." He would cause no trouble, meekly "play the game according to their rules." But all the while, he, shrewd and on guard, was watching and listening. All the while, he later boasted, he was making his way "further and further into the truly bizarre intelligence labyrinth that surrounds the subject of UFOs."

He suppressed all his natural emotions: his anger and revulsion at Bennewitz's torture, his impatience with the Falcon's capriciousness, his eagerness to run to the media for help and protection. He let everything well up inside. He held it all back, a fair trade, he felt, for his chance *to learn the truth*.

From the moment of his initial encounter with the Falcon, he had anticipated traps, and he was not disappointed. The first one was set early on, not long after he had been recruited to report on Bennewitz. The Falcon himself waved the bait.

"The rumors are true," Moore remembered the Falcon piously telling him. "We did shoot down a flying saucer."

Moore, a model of self-discipline, only uttered a stoic, noncommittal, "Yes."

But all the while his heart was leaping.

Then the Falcon, as if making idle conversation, proceeded to reel off the precise location of the encounter, the date, and the names of the military units involved. "Check it out if you don't believe me," the Falcon said, again as if an afterthought, before steering the conversation into less sensational arenas.

Moore did check it out. Three weeks later he told the Falcon what he had found. "No way that ever happened," he declared. There was no malice in his voice; his manner was crisp and professional. Moore, reading from his pages of notes, explained to the Falcon that the military units he had credited with having taken part in the engagement were not in the Southwest on the day in question. "Your information's phony."

The Falcon beamed. "Congratulations," he said. "You didn't run to the *National Enquirer*. You did what we hoped you would do. You passed the test."

Still, there was no diploma, only more tests. "Would you like to read a transcript of an interview with a captured extraterrestrial?" the Falcon coyly asked. Moore had been meeting with his control for over a year by now and he felt the relationship was solidifying. He was hopeful and, conforming to his cover, humbly curious.

He didn't protest, then, when the Falcon explained that a viewing of the transcript required some elaborate security procedures. Moore was ordered to fly from airport to airport across the country; at each stop, he was to hurry to a pay phone to receive his next destination. Eventually, after an exhausting and perplexing day in the air, he was directed to a motel in upstate New York. He was to sit in his room and wait for a call.

The call came, and so did further instructions. Moore was to go to the restaurant across the street. Take a window seat, he was told. You will be contacted.

But he wasn't.

Moore, though, didn't complain. And, either as rewards or simply as inducements to keep him eager, occasional intelligence crumbs continued to be thrown his way. For example, Moore said he was called on the phone in the fall of 1983 and told that Korean Air Lines Flight 007 "had been shot down over the Soviet Union—before the story hit the press." Other times, he was escorted into high-security areas inside government compounds. And, another hint of a promising future, he was introduced to an expanding circle of government intelligence operatives. There were ten in all, representing a variety of official agencies. He called them, in deference to the Falcon, his first contact and his first stab at word code, "the aviary." He no longer had any doubts about their position, their power. He wanted their help.

Yet Moore continued to stay on guard. He was being drawn into a network that, even if all its claims didn't prove true, was certainly dangerous; Bennewitz, broken and ruined, was proof enough of that. He needed backup, and he needed protection. Moore decided that he, too, needed a team.

He chose carefully. He wanted people he trusted, and people whose skills complimented his own. He was living in Los Angeles in June 1982 when he made up his mind to approach Jaime Shandera.

Shandera was no true believer. If he ever thought of UFOs, it was a rare day, and then only in the prosaic terms of his chosen trade—as a possible "project." Shandera was by profession a producer. At forty-five, he had knocked around the business for most of his adult life and had managed to earn a string of middling impressive credits working on documentaries for Time-Life Broadcast and RKO General. His was a professional life spent looking for the next deal, scurrying after the rights to the next hot property. In fact, that was how he and Moore had met. They had discussed at some length a fictional film inspired by Moore and Stanton Friedman's continuing research into the Roswell crash.

The film was short-lived; it was abandoned, no rare occurrence in fickle Hollywood, before it had even progressed to such plateaus as deal memos and treatments. Moore and Shandera's friendship, however, survived the project's demise. So when looking to put together his team, Moore's thoughts naturally turned to a trustworthy buddy.

Another advantage was Shandera's background in documentaries. His well-developed skills in sorting through complicated stories, Moore reasoned, would be bound to come in handy. Who knew, after all, what kind of traps, what kind of convoluted disinformation the aviary was going to throw at him next? It would certainly be a plus to hold these up to the light of Shandera's veteran judgment. Also, people liked him. Shandera had a producer's easy, backslapping friendliness. He was by instinct cheerful and outgoing, a man who liked to talk to people and who, in turn, most people enjoyed talking to. Moore, a terse, gloomy presence with a bouncer's menace, realized his limitations. Some people, he admitted to himself, might find Shandera more approachable. And that couldn't hurt.

So Shandera was recruited. "What producer could turn down a story like this?" he would later say, with a flashy, ever-ready smile. But on another occasion Shandera would reveal a harder side to his commitment: "I told Bill I would go the distance—win, lose, or draw. What I see here is a situation analogous to Watergate. We have to become involved in getting the real story at all costs."

And with Shandera signed up, Moore decided the time had also come to make Stanton Friedman, the nuclear physicist who was his longtime research associate, somewhat aware (or "semiconscious," as the spooks would discreetly say) of the game that was at hand. Friedman was living now in New Brunswick, Canada, away from the front lines, but he quickly volunteered his support.

Moore now had his team.

He was ready. He had passed a battery of demanding tests. He had obeyed all the aviary's commands. He had served them loyally, not betraying their confidences. Yet, he still had nothing to show for his nearly four years of pious cooperation. Who could blame him if he grew anxious, if his studious patience was finally wearing thin? But Moore, as both Shandera and Friedman would testify, showed no outward signs of uneasiness, of doubt. He remained convinced he had "secured a seat in the right theater." All he could do now, he understood, was to wait for the curtain to go up and the show to begin.

Yet when the curtain finally rose, it was Shandera who was sitting front row center.

Showtime was shortly after noon, on the eleventh of December, 1984. Shandera would be sure of the time because he had just gotten off the phone with Moore. They had agreed to meet in an hour at the Villa Sunset, a dingy spaghetti parlor whose main charm, besides, of course, the appealing prices on its tomato-stained menus, was its proximity to Shandera's house. It was only a five-minute drive away, so Shandera returned to his favorite chair and picked up his copy of *Variety*. He was reading about the progress of General William Westmoreland's libel case against CBS, a matter of real interest to Shandera, a man who earned his living producing documentaries, when he heard the distinctive creak of the screen door. He paid it little mind, though, and continued to read with some concentration. Still, some corner of his consciousness focused on the rough sound of something, bulky apparently, being wedged through the front-door mail slot. And just an instant later, he couldn't help but hear the soft thud as this parcel fell to the entry hall floor. He finished the article before putting down his *Variety* and going off to investigate.

Someone's returned my wallet! That was his first thought as he picked up the brown 8½-by-11-inch envelope from the hallway floor. He had lost his wallet a couple of weeks back, and in that first instant, the envelope in his hand, Shandera decided that some Good Samaritan had found it and put it in the mail. Clearly, someone had gone to a lot of trouble to make sure the package arrived safe and sound. The envelope was protected by a border of brown tape that ran along its seams and the address label was faultlessly typed. He looked for a return address, but there was none. So much for the Good Samaritan, he reconsidered swiftly. More likely the thief was returning the wallet after picking it bare. Perhaps he had a pang of conscience. Or maybe he just wanted to get rid of the evidence.

Whatever, Shandera told himself as he took the envelope into the living room. He was simply glad to have his wallet back. He busied himself opening the envelope, and what with all the brown tape it was quite a production.

Inside the envelope, he found another envelope. It was brown, too, and its edges were similarly sealed with tape.

And all at once Shandera knew this had nothing to do with his wallet. *It had come. The waiting was over.*

He moved quickly now, suddenly in a hurry, as he tore the second envelope open. Inside was a third envelope, a white one. It was like one of those children's games, he thought: one doll hidden inside another, and still another doll inside the next. But all the time he kept on peeling the tape off the envelope, eager to find the final prize.

Inside the last envelope was a small canister. And inside the canister was a spool of film.

Shandera studied the film carefully. It was a roll of Tri-ex 35 millimeter. It was black and white, exposed but undeveloped. He quickly put the film back into the canister, the canister back into the largest of the three envelopes, and then, the precious package tightly in his hand, he rushed off to meet Moore at the Villa Sunset.

Moore was already seated at their usual booth in the rear, working on a beer. He never finished it.

Bill Moore used a kitchen glass to measure the developer. He didn't have much photographic flair, but when the two friends left the restaurant, it was agreed that it would be quicker—and more secure—if he developed the film himself. So now they were in Moore's house, and he was going through the trial and error of exposing the photographic paper for too long, then not long enough in the bathroom sink filled with developer. The next step, too, was a muddle. How long the exposed contact sheets should lie in the fixer was anybody's guess.

Still, when the images started to come into focus, there were shouts of triumph and Moore ordered Shandera to grab the contact sheets and follow him. They went to the living room and Moore, from God knows what drawer, produced a handful of clothespins. They used the clothespins to hang the contact sheets from a living room curtain rod. Moore opened up a window; the warm breeze would help the prints dry.

But even before the prints were dry, Moore, a magnifying glass in hand, was studying the contact sheets. He quickly decided it was a government document of some sort that had been photographed. Each frame revealed another typed page. Stamped on the top of all the pages, he noted with building excitement, was "Top Secret/Eyes Only." The typed words were legible and he began to read.

He was stunned. This was how he first came to see the MJ-12 documents.

As Moore finished one frame and moved on to the next, he felt a rush of victory, a sudden understanding that the four years he had put in with the Falcon had not been in vain. His version of the Roswell incident had been verified. Shandera hovered next to him, leaning in close toward the prints; there was only one magnifying glass and Moore wouldn't think of relinquishing it. Shandera, squinting, his nose nearly touching the paper, had a harder time deciphering the documents, but he was no less compelled to keep on reading.

When they were both done, the damp contact sheets hanging limply from the curtain rod like drying underwear, Moore was the first to speak.

"Are these for real?" he asked.

THIRTY-FIVE.

Phil Klass shared the same concern. It was May 1987, two and a half years after Moore had first uttered his bewildered question. But, with a nice touch of symmetry, Klass's initial reaction to the MJ-12 documents was near enough to be an echo: Are these for real?

In his heart, he knew they weren't. But prejudice is not proof; so Klass, in an act of rigid self-control, reined in his galloping suspicions. With a reporter's cool objectivity, he began to review the big pieces of the puzzle as he understood them.

Fact: In 1980 Bill Moore publishes a book that insists a flying saucer had crashed near Roswell, New Mexico, in 1947, that alien bodies had been found in the wreckage, and that the U.S. government had done its best to cover up this historic event.

Fact: Four years later, an associate of Moore's, television producer Jaime Shandera, receives a roll of exposed film in the mail from an anonymous source. Moore himself develops the film—and the prints reveal allegedly top-secret government documents.

Fact: These documents not only confirm all the claims of Moore's book, but also detail the formation of a panel of government officials who will deal covertly with this alien threat.

Fact: Moore, Shandera, and Friedman decide to keep silent about the existence of the MJ-12 documents while they work to establish their authenticity. After two years of research, the trio decides to go public. "Based upon research and interviews conducted thus far," their press release proclaimed, ". . . the document and its contents *appear* to be genuine. . . ."

But Klass, these facts now lined up in his mind as though they were dominoes waiting to be felled with a single, soft push, wanted to scream that appearance was a world apart from reality. The whole series of events, he felt, was too neat, too coincidental. Moore writes a farfetched book and, like an answered prayer, he finds top-secret papers that validate his wild story! Lady Luck was sure smiling on Bill Moore that day; and Klass, by instinct, had little faith in the likelihood of such acts of convenient benevolence.

Instead, from the start, he had a thousand questions. Why had the film been sent to Shandera, who had never published any books or papers on UFOs? How did the sender know the film would find its way to Moore? Why did Moore decide to develop the film himself? Or, just as puzzling, why had the sender, after apparently going to so much trouble, not taken things one step further and processed the film before passing it on? And after merely skimming the papers, his questions multiplied. Why was the briefing document written as if President-elect Eisenhower knew nothing about the crashed saucer recovered by Army officers when, in 1947, Eisenhower had been Army chief of staff? Why did the document cite the "upsurge" in UFO sightings at the end of World War II, but fail to mention the more timely and more menacing sighting scares in the vicinity of the White House in July 1952? Why was the Truman memorandum written in such an un-Trumanlike, convoluted, bureaucratic style?

And so on. His mind bubbled with questions, and suspicions.

But enough! Klass realized he was getting ahead of himself. His assault had to be orderly. First he would analyze the evidence in detail, then he would ask the questions. He turned once more to the MJ-12 documents themselves. They had to contain, he just knew, the incriminating evidence. He read them over and over and over. In his mind he might just as well have been a dogged detective, the sleuth who keeps on scouring the scene of the crime confident that eventually he will stumble upon a previously ignored clue—the well-hidden smoking gun.

The names were not listed in alphabetical order. Nor was the chairman of the group identified. In fact, as best he could determine, there was no logical sequence to the list. Those were Klass's first observations as he studied the Majestic-12 Group membership roll that was on the first page of the alleged Eisenhower briefing paper. Then, neither having scored a quick victory nor found any reason to admit defeat, he abruptly transferred his attention to the twelve names—which was where, Klass would later say, he had been eager to go from the start.

ADM. ROSCOE H. HILLENKOETTER. Hillenkoetter was the first director of the CIA. A logical choice for the group, Klass noted. But it didn't take much musing before he decided that a shrewd counterfeiter would also have included Hillenkoetter because of the admiral's well-known involvement with the UFO movement. After he had retired from the Navy, Hillenkoetter was quoted in *The New York Times* as saying, "It is time for the truth [about UFOs] to be brought out in Congressional hearings." A nice touch, Klass thought. The counterfeiter wanted people to believe that Hillenkoetter actually knew about a crashed saucer and that was why he went public in his old age. Perhaps a little *too* nice. And something else, after a bit more digging, gave Klass further pause. Authentic letters obtained from the Truman Library showed that Hillenkoetter used "R.H.," his initials, not "Roscoe." Also, these genuine letters correctly indicated his rank as "Rear Admiral," not the four-star full "Admiral" implied in the briefing paper.

DR. VANNEVAR BUSH. One of the country's leading scientists, Bush had organized the National Defense Research Council and, in 1943, the Office of Scientific Research and Development which led to the establishment of the Manhattan Project. A definite candidate for any government group looking into flying saucers, Klass had to admit. But Klass also learned that Bush had resigned from the Defense Research and Development Board in 1948. It was his wish, Bush had written Truman, "ultimately to be free of governmental duties in order to return more completely to scientific matters." Yet the MJ-12 documents claimed that four years later Bush was still actively involved with a government panel. Curious, Klass decided.

SECY. JAMES V. FORRESTAL. Forrestal, too, as secretary of defense, was another prime recruit for the group. Klass found no way to challenge that possibility. A footnote to the document, however, raised some questions. It stated that following Forrestal's death in 1949, more than a year passed before his place in the MJ-12 group was filled by General Walter Bedell Smith. And that bothered Klass. The new defense secretary, Louis A. Johnson, should have been Forrestal's logical and immediate replacement. And Smith, who became head of the CIA in 1950, would have been Hillenkoetter's MJ-12 replacement when the admiral returned to active duty in the Pacific as commander of the Seventh Fleet.

GEN. NATHAN F. TWINING. Twining was commander of the Air

Materiel Command at Wright Field (now Wright Patterson AFB, Dayton, Ohio), where, according to the MJ-12 documents, the crashed saucer was sent for analysis. But, Klass discovered, within weeks of the supposed creation of MJ-12, Twining was transferred to head the Alaskan Command. And that didn't make any sense. Why wasn't Twining ordered to remain at Wright Field to supervise and report on the examination of the crashed saucer? Better yet, why wasn't his successor at Wright Field named to replace him on MJ-12?

GEN. HOYT S. VANDENBERG. Vandenberg was Air Force chief of staff. Naturally, any counterfeiter would smugly believe he could get away with putting the general on the MJ-12 panel, Klass reasoned. There was, though, one problem raised by Vandenberg's alleged participation. If the head of the Air Force knew UFOs were real and a threat, why had the Air Force failed to take seriously the mysterious (and later rationally explained) blips that were observed on the radar at Washington's National Airport in 1952? Wouldn't the general have taken some genuine precautions if he had reason to believe the intruders might also be genuine?

DR. DETLEV BRONK. Another safe choice. Bronk was an internationally respected physiologist and biophysicist. He was chairman of the National Research Council, a member of the Medical Advisory Board of the Atomic Energy Commission, and (with Dr. Edward Condon, who later headed the Air Force–sponsored Project Blue Book) a member of the Scientific Advisory Committee of the Brookhaven National Laboratory.

DR. JEROME HUNSAKER. Hunsaker was an aircraft designer, head of MIT's aeronautical engineering department, and chairman of the National Advisory Committee for Aeronautics. If someone had been chosen in 1947 to examine a crashed saucer, Hunsaker, Klass had to admit with a sigh of surrender, might very well have been the man.

MR. SIDNEY W. SOUERS. Souers, the first director of Central Intelligence, was executive secretary of the President's newly created National Security Council in 1947. But he retired in 1950 and returned to civilian life. Why, Klass wondered, was he listed in late 1952 as still being an active MJ-12 member? Why wasn't his successor at the NSC also taking his place in the Majestic Group?

MR. GORDON GRAY. A truly unlikely candidate, Klass ruled after a quick review of Gray's career. Gray had been *assistant* secretary of the army when MJ-12 was created; he didn't become secretary until two years later. Besides, his background revealed no immediate skills that he might have offered the group. Gray had been trained as a lawyer, had spent the previous ten years as the publisher of two newspapers, and he did not hold a prominent Pentagon job.

Dr. Donald Menzel. Another respected scientist, Menzel was a director of Harvard College Observatory. His expertise was in solar and stellar astronomy, but Klass, now fuming, reached a more devious theory as to why the Majestic Group had included Menzel—punishment. Menzel was a well-known UFO debunker. The author of three books on the subject, his general position was "All non-explained sightings are from poor observers." And Klass (who knew a thing or two himself about the vengeful ways of the true believers) was certain Menzel's inclusion in the MJ-12 group was one more snide shot at the astronomer: Menzel was not merely a foolish debunker, but, worse, he had knowingly lied.

Gen. Robert M. Montague. He was base commander at the Atomic Energy Commission installation at Sandia Base, Albuquerque, New Mexico, from July 1947 to February 1951. If an alien ship had crashed near Roswell, he undoubtedly would have known, Klass conceded.

Dr. Lloyd V. Berkner. The last man on the list, Berkner was also presumably picked for his scientific expertise. He was executive secretary of the Joint Research and Development Board, and he had headed a special committee that led to the establishment of the Weapons Systems Evaluation Group. Solid credentials: another logical choice. Or, Klass reasoned after some ferreting through the library, maybe not. Berkner had also been a member of the secret CIA panel convened by Dr. H. P. Robertson to study the Air Force's most likely UFO cases. The Robertson Panel's final report, classified "Secret" and signed by Berkner in 1953, had determined there was no evidence to support the theory that UFOs were extraterrestrial crafts. Which left Klass asking, why would Berkner, a busy scientist, have participated in another time-consuming study, as well as endorse its dismissive findings, if he, an MJ-12 member, already knew the real story?

And, as Klass scanned the list one final time, he realized there was at least one thing the twelve MJ-12 members undeniably had in common—they all were dead. Which was most unfortunate.

Then again, it was also quite convenient. Dead men don't tell tales. Or debunk them.

. . .

All of which struck Stanton Friedman as completely absurd. It wasn't simply because he was Bill Moore's friend and research associate. Rather, he realized what was at stake. The MJ-12 documents were an opportunity to prove once and for all that UFOs were real. So he, a judge climbing to the bench in an admittedly more sympathetic court, listened to Klass's objections and bellowed, Overruled. Don't tell me Ike should have been informed about the saucer immediately after the crash. A month earlier he had announced his plans to leave the Army to become president of Columbia University: he no longer had any "need to know." As for the briefing itself, a *Washington Post* article confirmed that President-elect Eisenhower had indeed received a military briefing on November 18, 1952—the very day specified in the MJ-12 documents. Klass's other concerns, Friedman insisted with thunderous vehemence, were equally contrived. Documents from the Truman Library, for example, repeatedly referred to the first CIA director in a variety of ways: Admiral Hillenkoetter and Rear Admiral Roscoe H. Hillenkoetter, as well as other variations. Not simply R. H. Hillenkoetter, as Klass had argued. And did Klass seriously believe that simply because General Twining had been transferred to Alaska he was out of touch with MJ-12 activities? "We're not talking about privates here who have to rely on the commercial mail system," he sneered. And for an added confirming touch, Friedman gleefully waved about a letter written by Twining just ten days after the Roswell incident. The general was canceling a long-planned trip to see a prototype Boeing airplane due "to a very important and sudden matter that developed here." His new destination—New Mexico.

But Friedman, too, despite his every inclination, didn't rush to judgment. He didn't immediately proclaim the MJ-12 papers as genuine beyond a shadow of every reasonable skeptic's doubt. Instead, his mind, too, was restless. And like Klass, he proceeded methodically. He also began by spending many speculative hours studying the list of MJ-12 members.

And from *his* perspective, based on *his* findings, the names only served to help authenticate the documents. All the men listed were individuals who would logically have been selected in 1947 to serve on a top-secret government panel to investigate a crashed saucer.

All except one—Dr. Donald H. Menzel.

That appointment troubled Friedman. There was no apparent reason for Menzel's traveling in such auspicious circles. Why should a professor of astronomy be selected to be part of a high-level military and intelligence cover-up?

Unless Klass had been correct—a counterfeiter was simply trying to have a vindictive last laugh on Harvard's great debunker.

Perhaps if Friedman had not so disliked Donald Menzel, he might have been prepared to let the question linger unresolved. But, as Friedman would be the first to admit, he had an ax to grind. "Here was," he would explain, all at once feisty and bitter, "a famous Harvard astronomer whose reasoning about UFO sightings could be picked to pieces by a college freshman physics major." And no doubt a greater sin in Friedman's world, Menzel's "first book in 1953 was translated into Russian and kept almost a whole generation of scientists in both the U.S.S.R. and the U.S.A. from even nibbling at the UFO phenomenon." So it was with a sense of curiosity and anticipation that was nothing short of personal—in addition, of course, to his unwavering professional commitment to help Moore and Shandera get to the bottom of things—that Friedman left his home in New Brunswick, Canada, and headed for the musky archives of Harvard University in Cambridge, Massachusetts.

Once at Harvard, Friedman, to his considerable embarrassment, had to spin a story about the paper he was planning to write on public-spirited postwar scientists (which was true—to a point) before he was allowed access to the Menzel archives. And, more painful, he then had to appear before Mrs. Menzel, a gracious and gray-haired widow. Once again, as she smiled unsuspectingly at the visiting scholar, he had to trot out the same thin tale before she decided that yes, it would be perfectly proper to allow Mr. Friedman to read her husband's unpublished autobiography. But it was worth it. Small deceits were quickly forgotten as soon as Friedman began to burrow through the unpublished papers. For what he found took him, he exclaimed, "totally by surprise"—Donald Menzel had led two lives.

The public one, Friedman wrote, "was as a famous astronomy professor serving on eclipse expeditions, establishing solar observatories, supervising graduate students, and debunking UFOs." The second life was clandestine. Menzel was a consultant for the CIA and the NSA with a Top Secret Ultra clearance, he had worked on classified research projects for more than thirty industrial companies, he had made frequent trips to Washington, D.C., and New Mexico on "government business," he was a close associate of Vannevar Bush, Lloyd Berkner, and Detlev Bronk, and he was a man who had spent his adult life keeping these secrets.

Menzel was certainly a logical candidate to have been chosen for the Majestic-12 Group.

Now with a sudden clarity, Friedman realized his decades of animosity toward Menzel had been ill-spent. In fact, the professor was a man worthy of respect. Menzel had been a patriot following his country's orders. Friedman now understood that Menzel's three books on UFOs were poorly researched *by design*. Menzel's role on MJ-12 was to spread disinformation about flying saucers, to help keep the secret.

But it was too late for apologies, and besides there were more pressing concerns. It was with a sense of great triumph that Friedman posed a question that allowed him to fit another piece into the puzzle. How could a casual forger have known of Menzel's close association with intelligence agencies? And Friedman answered his own question with a confident shout. A forger couldn't; the connection was unknown to the public until Friedman's research. A counterfeiter would never have designated Menzel to serve on the panel.

But—and this was Friedman's great deduction—President Truman might very well have picked the Harvard astronomer. The President would have known Menzel had all the correct qualifications.

Dead men, Friedman now rushed to correct Klass, *do* tell tales.

THIRTY-SIX.

"It proves nothing," Klass explained with the elaborate weariness that was part of his debater's style when he learned of Friedman's discovery in the Harvard Archives. "So what if Menzel could have been a member of MJ-12? That doesn't prove the man *was*."

To which Friedman, forever the bulldog, barked back, "He's avoiding the issue."

And that, to a degree, was true. Klass, now that he was caught up in the complex task of searching through the text of the documents—"The papers are the only clues we've got," he kept on insisting—was suddenly reluctant to head off in pursuit of diversions. To his mind, chasing after Friedman's challenge would be a fool's game. It was a match that, at best, could only end in a draw: Even if Menzel had been a master spy, the revelation would, ultimately, neither prove nor disprove the MJ-12 papers' authenticity. Klass wanted more. His research strategy was geared for nothing less than total victory. He needed to demonstrate that UFOs did not exist. Doubter and no less avenger, he was not about to allow anyone to divert the constant thrust of his determination. He remained confident he would find the counterfeiter's sloppy fingerprints right before his eyes—if he continued to look hard enough.

And still he nearly missed them. They had been, he would decide, staring at him from almost every page, just as he had suspected all along. But it took him a while to realize they were indeed clues; and even longer to follow their speculative path to a larger, more incriminating theory.

His breakthrough came without warning. Early on in his research, Klass had made a list of all the dates contained in the MJ-12 pages.

It was a simple exercise. He had written the numbers in a column down the left-hand margin of a pad like a schoolboy preparing for a history exam. And then, moving down the list, he had begun to ask himself questions about each specific date. Could, for example, President-elect Eisenhower have conceivably been briefed on November 18, 1952, as the document claimed? It was a way of organizing his research, and it was demanding. He had already spent many long weeks absorbed in the struggle to answer all the questions each specific date raised.

Until it came to him. Out of the blue. He was looking at his pad one moment and then at the MJ-12 documents themselves the next, and all at once he noticed that the two sets of dates were *different*. No, not the specific dates, he corrected himself; rather, the way they were written. The dates on his pad were listed in the traditional manner. He had written, he saw: November 18, 1952. But the MJ-12 briefing papers had used a peculiar way for writing dates—an eccentric mixture of civil and military styles. Using the standard military format one would have written 18 November 1952. But the briefing document had been typed by someone using a military form with an unnecessary and erroneous comma: 18 November, 1952.

He began flipping through the papers, quickly scanning each page. And there it was, on page after page. Every date in the documents had this "unnecessary comma." It was an error, Klass was certain, that a government typist would never have made.

Now on a roll, excited by the renewed thrill of being back in the heat of the chase, he started in once more at page one and soon discovered something else. There was a zero preceding every single-digit date in the document. The briefing pages contained such anomalies as "01 August, 1950" and "07 July, 1947" and "06 December, 1950." He had spent much of his professional life reviewing government papers and he knew that the official style, at least until the computer age, would have been to type 1 August 1950. It was another odd error, one a government typist would never have made.

And then he suddenly remembered. It was more than odd.

It was familiar.

He hurried to his personal files. Rapidly, feeling that his long quest was perhaps nearing an end, he began sorting through the folders until he came to the one with Bill Moore's name on it. Picking up a sheet at random, he found on his very first attempt what he had recalled—

the date on a letter he had received from William L. Moore Publications & Research. It was typed with an "unnecessary comma" as well as a zero preceding the single-digit date.

Later, after a more exhaustive search, he would write:

"By a curious coincidence, this [the unnecessary comma] is precisely the same style used by William L. Moore in *all* of his many letters to me since 1982. . . .

"My files of correspondence from Moore show that he used a single digit *without* a zero until the fall of 1983—roughly a year before the Hillenkoetter document film reportedly was sent to Shandera—when he then switched to the same style used in the Hillenkoetter briefing document."

Klass's theory was carefully expressed, his words chosen to suggest rather than to state. But for all his precautions, Klass might just as well have screamed *Gotcha!*

Which left the ball in Bill Moore's court.

The date format was, he was quite ready to concede, "the most controversial aspect of the MJ-12 document affair." Yet when he finally decided to confront the issue publicly, he, protective of his role as the man on the inside, chose to make remarks that, at first, mixed an ingratiating candor with an equally infuriating obliqueness. But as he spoke, the combative side of his complicated personality began to break out. It wasn't long before he let loose with a welled-up, snarling, even baiting attack. His arena was the relatively friendly turf of the Mutual UFO Network's (MUFON) annual conference in Las Vegas, and he was nearing the end of a lengthy speech when he, pausing first to take a gulp of air as if for sustenance, turned without warning to the provocative questions Klass had raised. He began, "There are a couple of points I would like to make." But by the time he was done he had exceeded his self-prescribed limit; and, no small comfort, the audience was applauding madly:

"One. If similarities between stylistic characteristics of the documents and my own writings are important, then are not dissimilarities equally important? There are some, you know—even in the way the dates are written. Why do those who go to such pains to point out similarities not go to equal pains to point out dissimilarities? Could it,

I wonder, be prejudice? Or does the word 'incompetence' better describe it?

"Two. We are by no means convinced that the date format is as critical an issue as others would make it out to be. While the style is admittedly unusual, it is not without precedent.

"Three. To those of you who still harbor suspicions that I created the documents and am perpetrating a hoax, I say, 'Thank you for the compliment!' To those of you who think it more likely that I am the innocent victim of an elaborate hoax, I concede that remains a possibility which has not yet been entirely ruled out."

Phil Klass, sitting toward the rear of the large room, the self-proclaimed "skunk at the garden party," a tape recorder balanced on his lap, heard it all. To his sharp ears, there was nothing worth applauding. Moore's explanation was "evasive at best." "He's trying to have his cake and eat it, too," Klass complained.

Later that summer, now back home in Washington, he would play the tape of Moore's speech over and over. There was an unmistakably feisty quality to Moore's words, yet Klass would never fail to be charmed by the incongruously soft, woolly murmur in his enemy's timbre. Moore had a lovely speaking voice. Klass, an orderly man, would always start to play back the speech at the point of Moore's opening words. But it wouldn't take long before he grew impatient and would fast-forward precisely to Moore's discussion of the dates. "One," he would hear Moore begin, stressing the syllable like a referee giving the count; and, in an unconscious reaction, Klass would immediately move in closer to the machine. He was fascinated—and enraged—by what he had come to call "Moore's half-hearted argument." And by the time Moore's sly, teasing voice had facetiously thanked all those who believed he was capable of pulling off a hoax like the MJ-12 documents, Klass could not help but loudly reply, "You're welcome."

While he silently thought—case closed. I have proven once again that UFOs are not real.

THIRTY-SEVEN.

But it was not that easy. For just as Klass was poised to vent his private thoughts, just as he was ready to bring the gavel down on the MJ-12 documents and declare them once and for all an indisputable fraud, his opponents, returning to the fray with an almost uncanny sense of the moment, would time after time find one more provocative point to bolster their case. Friedman was a master; no researcher could have been more industrious, no lawyer could have mounted a more aggressive rebuttal. What's this about the dates? he would rumble. Klass's quibbles prove nothing except typically thick-headed "Klassical reasoning." And then Friedman, playing his trump cards with flair, would produce not only a small stack of government and military documents with "unnecessary commas" and zero-prefix dates, but also a letter written by none other than Admiral Roscoe H. Hillenkoetter himself which was dated, for all the world to see, with a resplendent "unnecessary comma." Take that, Mr. Debunker. UFOs were real.

Still, it was Moore and Shandera, a team that never failed to be blessed by Providence, who uncovered what many felt was the most convincing piece of evidence to support the authenticity of the MJ-12 documents. Only this time, the mailman didn't deliver a roll of film—just a perplexing clue that, when solved, sent them off on a most rewarding treasure hunt.

The hunt began with a postcard. It was addressed to the post office box Bill Moore still maintained in Dewey, Arizona, and then forwarded to his new home in Los Angeles. It was unsigned, and, it seemed, a bit of care had been further spent to protect the identity of the sender. The card had been postmarked in New Zealand; however, the photograph on the front, credited to the Ethiopian Tourist Commission, displayed a colorful scene from the African bush. But it was the card's message, single-spaced and typed with precision, that both Moore and Shandera decided was the more pressing mystery. It appeared to be a puzzle of some sort.

To this day, neither Moore nor Shandera will reveal the entire message typed on the postcard; they are men who cling protectively to their hard-won secrets. After much urging, however, they will share a bit of the card's riddle. "To win the war," the card promisingly began. And it ended:

Add zest to your trip to Washington
Try Reeses pieces;
For a stylish look
Try Suit Land.

The card had both men stumped. "To win the war": that was the easy part, Moore immediately decided. The "war" had to be the battle of a lifetime he was presently waging—his all-out attempt to prove the MJ-12 papers were genuine: that an alien spaceship *had* crashed in New Mexico, that UFOs *did* exist. But what did the rest mean? Was it really a signpost that would guide him to the final victory? Or was it more disinformation? Another false promise?

Shandera did not think so. He, the film buff, after much deep thought, came up with the beginnings of a solution. Reese's Pieces—and he was quick to dismiss the missing apostrophe as an unintentional error—reminded him of the movie *E.T.* After all, wasn't that kid in the movie, Eliot, always gobbling down Reese's peanut butter and chocolate candies? Clearly, he reasoned, the riddle had something to do with extraterrestrials. But try as Shandera might, he couldn't take his theory any further. "Suit Land": that really had him baffled. Maybe, he finally suggested to Moore, it was a reference to the clothes the extraterrestrials had been wearing when their ship crashed near Roswell. Perhaps the message when solved would lead them to actual outfits from outer space?

Could be, Moore acquiesced. But he was not very convinced. And it did not help his mood that, despite hours of deep and concentrated thought, the code on the card was still unbroken. It remained one more teasing mystery in a life that, since his first conversation with the Falcon, seemed determined to lead him deeper and deeper into clandestine shadows.

But as things worked out, it was Stanton Friedman's turn once again to provide the missing link—only this time he didn't even know he was coming to his friends' rescue.

It all began after Friedman had a telephone conversation with JoAnn Williamson, chief of the military reference branch of the National Archives. He wanted to know the status of Record Group 341, a vast collection of top-secret Air Force Intelligence files that was in the process of being declassified. There were so many individual papers involved that six four-person teams of Air Force classification specialists had been working for the past three months to complete their review. And at least once a week throughout those three months, Friedman had been phoning Ms. Williamson to learn if the papers were, at last, available to the public.

That day he was told they were.

Friedman was elated. He wanted very much to believe the newly declassified files contained evidence that would reinforce his belief in the MJ-12 documents. Yet, just as he began to plan his excursion to the National Archives in his mind, his mood went into a sudden skid. He abruptly realized there was a problem: he couldn't go. He was scheduled to embark on a lengthy and hectic UFO lecture tour. It was a financial necessity; it couldn't be postponed. He moped around for a while, and finally called Bill Moore in Los Angeles.

Moore, as Friedman would remember the conversation, was not very sympathetic. It was a gloomy period in his life; the unfulfilled promise of the postcard had left him with little hope of ever winning the war. Glumly, he began to lecture Friedman that he really didn't see the point in anyone's rushing off on another wild-goose chase. He, for one, was certainly not willing to drop everything and head off to the National Archives in Washington.

Actually, Friedman explained, the 341 files weren't in the Archives' main facility in Washington. Like most of the more recently declassified military papers, they were stored in a satellite branch. In Maryland.

"Maryland," Moore would recall repeating with some exasperation.

"Yes," Friedman explained patiently, "Suitland, Maryland."

And if Moore still had any small doubts about the solution to the postcard's riddle, they were assuaged when Friedman, baffled by his friend's sudden, new-found eagerness to look at the 341 files, happened to mention that it would be necessary to request them directly from the office of the military archivist in charge—a Mr. Edward Reese.

"How will we know what we're looking for?" Shandera had asked.

"When we find it, we'll know," Moore, Zenlike, had answered.

And so, determined, ready for the hunt, five days after Friedman's call, the gung-ho team of Moore and Shandera assembled at the Suitland archives. At first, it was great fun. They would claim a desk bright and early each morning and the archivist, on their signal, would roll over a metal cart loaded with two heavy cartons from Record Group 341. A box would be placed in front of each of them and their work would begin. Each box was a treasure trove of once-classified government documents. They would read all day, hardly stirring.

On the morning of day one, as they reached greedily into that first box and immediately extracted a handful of papers all stamped "Top Secret," there was an intoxicating sense of adventure. They were starting out on a journey to explore a long-forbidden land. But by the afternoon of the third long day, after reading through hundreds of buff-colored folders filled with either obscure or indecipherable or uninteresting memos, the task had become dreary. The room was stuffy, the air-conditioning barely a weak breeze in the middle of a hot summer, and the anemic glow escaping from the tinted bonnets of the desk lamps only made the long and laborious hours of reading more of a challenge. They had worked their way through 125 heavy cartons, and by now it was not hard to wonder if this, too, was nothing more than one of the Falcon's demanding tests. But they asked for still another carton.

The archivist, pushing the metal cart as lackadaisically as any shopper moving down the aisle of a Stop 'N Shop, approached; and the

126th carton—Box Number 189 was its official designation—was placed in front of them. Moore and Shandera dug in.

Shandera was skimming the papers in folder T4-1846, his eyes nearly bleary, when something he was reading all of a sudden filled him with a rush of discovery. With great care, more an odd reluctance to recognize what he held in his hand than an act of scrupulousness, he began to reexamine the page. It was an unsigned carbon copy of a brief memo, dated July 14, 1954, from Robert Cutler, special assistant to President Eisenhower, to General Nathan Twining. Its subject—"NSC/MJ-12 Special Studies Project." His heart racing, he would later admit, "like at opening night," Shandera reread the words that had so shaken him:

"The President has decided that the MJ-12/SSP briefing should take place *during* the already scheduled White House meeting of July 16 rather than following it as previously intended. More precise arrangements will be explained to you upon arrival. Please alter your plans accordingly.

"Your concurrence in the above change of arrangements is assumed."

Shandera passed the page to Moore with only one soft yet succinct comment—"Bingo!"

Later a triumphant Moore would announce that the Cutler-Twining memo found at the National Archives "unquestionably verifies the existence of an 'MJ-12' group in 1954 and definitely links both the National Security Council and the President of the United States to it." And, as he explained to his supporters, since the Majestic Group was real, then so were UFOs!

But if Shandera couldn't believe his luck, Klass wouldn't. It seemed, even in the most munificent realm of possibilities, too coincidental. Moore and Shandera travel across the country, roll up their sleeves, and dig into a mountain of documents never before examined by any civilian researchers, and after a mere three days find *precisely* the proof they need to authenticate the existence of the MJ-12 Group. It was, he would explain with arch politeness, "most unlikely." Besides, he quickly pointed out after consulting authorities at the National Archives, all the other top-secret documents in the Record Group 341 boxes were stamped with an individual "register number"—an inventory code used by the Air Force declassification specialists. The document Moore and Shandera had found did not display such a number.

Someone, Klass decided, had planted a counterfeit memo in the archives.

Which Friedman, eager to get back into the match, conceded. "The memo might have been planted," he agreed. But, he went on, the key question was—Who planted it? It couldn't have been Moore or Shandera. The Suitland facility was carefully monitored; they weren't even allowed to bring pads and pens into the document review area. It was impossible that either of his two friends could have smuggled the document, found without an incriminating crease, into folder T4-1846. But he did have a theory. The Falcon and his furtive cohorts might very well have been able to pull off such an operation.

"Considering that the memo has no connection with any other items in Box 189 where it was found," he wrote in an issue of *International UFO Reporter*, "and that at least six four-person teams of USAF classification specialists were involved over a period of more than three months and were able to bring in briefcases and notebooks, it seems certainly possible that the Cutler-Twining memo was indeed planted to be found by Moore and Shandera at the instigation of the same group that had earlier leaked the film."

But, of course, Klass had no faith in the story about the film or the Falcon to begin with.

The long debate, a continuing succession of charges and well-timed countercharges, kept going back and forth—and still neither side could shout with total conviction, "Case closed." Even after two harsh years, the battle remained heated; both sides were committed to working to unearth facts that would support their arguments, and, no small corollary concern, their ways of looking at the world. It was their mutual obsession, and their shared duty. Moore, shaking with fury, would demand, "Why does Mr. Klass bash UFOs? I, for one, would welcome some rational explanation of just why a man would devote twenty years of his life to an essentially one-man crusade against something which he claims doesn't exist.'' And Klass, making a subtle point about his adversary in the process, would shift his usual reserve down one more restraining notch, and, his tone as steady as his conviction, say, "A twenty-year crusade for the truth doesn't seem like a folly to me.'' And so, each certain of what was truth and what was folly, they battled on.

THIRTY-EIGHT.

The meeting in the Tank was a war party. In the months that had followed Colonel Phillips's distribution of the Xeroxes of the MJ-12 documents, at least a few of the UFO Working Group members developed variable and complex suspicions. One moment they would be convinced the documents were an absurd fraud; while the next, players of the Great Game themselves, they would acknowledge that it was indeed possible that a large mystery was being protected by some level of the government just beyond their grasp. It was an insecure, unsteady time. And now, taking their seats around the walnut conference table on a fall morning in 1989, they wanted answers. Had a UFO crashed in New Mexico? Were aliens visiting this planet?

Colonel Phillips no doubt read their testy mood; and perhaps that was why he, a shrewd manipulator of his peers, decided to have the first word. This morning he was seated at the table, not standing up at the lectern (and this, too, was another wise touch; look, it said, I'm in the same boat you are), and before anyone else spoke, his concerns poured out in a rapid-fire torrent.

Why, the colonel asked, according to one person who listened attentively, would someone go to all the trouble it takes to counterfeit such a document? Consider, he said, what a hoaxer would need to know. He would have to be aware of the dates when meetings between the President and his advisers could conceivably have taken place and who would be logical members of such a committee, as well as the precise historical circumstances surrounding the alleged Roswell crash so that he could reasonably embellish it into a larger event. And, Colonel Phillips further pointed out, such knowledge, while no small task, was only background information. The counterfeiter would also need to make the document look real. He would have to be aware of the correct form of top-secret coding classification stamps, where these stamps should be positioned on each page, and standard government procedures for numbering the pages of classified reports. And he would also need to include such accurate stylistic touches as the inconsequential yet deliberate textual errors government typists are instructed to add to each unique classified version of a report, "errors" that help trace the culprit if the document is leaked to an unauthorized source.

The MJ-12 documents, the colonel announced, passed all these cursory tests. The papers, he emphasized, looked and sounded real. Even down to three apparently deliberate "errors."

Why, he repeated, would someone go to all this trouble to counterfeit documents? What would they have to gain? And who would have the resources and the knowledge to pull off such an intricate and convincing hoax? It was these two factors—motive and ability—that, the colonel authoritatively explained, were the focus of the FBI's investigation as it attempted to determine if the documents had been forged.

He was doing his best to move along quickly now, hardly referring to the yellow 3-by-5 cards he kept in a neat stack in front of him, but all at once, from all around the table, he was interrupted—"pelted" was the way the moment was actually remembered—by questions. The members of the Working Group wanted to know if the FBI had, in fact, concluded that the MJ-12 documents were counterfeit.

The moment was at hand, but it was not to be seized. The issue was finally defined, but the FBI, despite the colonel's strenuous efforts to plead its case, was quite clearly an organization ruled by the most political of philosophies—sophistry. He explained that the FBI had made its inquiry into the MJ-12 documents a Priority One investigation (which later led a member of the Working Group to worry about the fate of Priority Two cases). Agents from the Bureau's Foreign Counterintelligence squads on both coasts were involved in the inquiry. Special Agent Nicholas Boone, a much decorated veteran after eighteen years on the job, led one team out of the Los Angeles office; Special Agent William Zinnikas headed a similar squad that worked out of 23 Federal Plaza in lower Manhattan. After more than a year of inquiries, after months of showing the papers to officials representing each of the nation's intelligence organizations, the FBI could not find an agency or an individual willing to swear out a complaint asserting that the MJ-12 documents had been stolen from their classified files. Therefore, if the papers weren't stolen, they must be forgeries.

"I'd like to meet the G-man who swept the whole thing under the rug with that neat bit of logic," loudly complained one of the NSA officials serving on the Working Group.

The colonel, however, insisted that the FBI was still actively pursuing its investigation. It was plugging away to determine who would have both the motive and ability to manufacture the documents. So far, the Bureau had come up with three possible scenarios.

The first, the colonel explained, involved a foreign intelligence service. The Soviet, the Chinese, even the rather inventive Bulgarian spooks were all trotted out as the possible villains. Certainly, those services, as well as a handful of less likely candidates, had the ability to fabricate classified, official-looking documents. As for motive, the FBI thought that was readily apparent. Muddying the waters, creating dissension, spreading paranoia in the ranks—those were all the day-in, day-out jobs of the ruthless opposition. But, after some digging, the FBI had come up with a more specific reason for the enemy's putting such a complicated plot in motion—revenge. It seemed in the 1960s that the CIA had done its best—again for the usual Cold War reasons—to spread tall tales about menacing flying saucers throughout the Soviet Union and in the southern provinces of the People's Republic of China. Now Moscow Center or the Chinese were having their turn with an ingenious bit of one-upmanship. Or at least that was one theory the FBI was busily pursuing.

Next on the FBI's list of suspects was the diligent band of Air Force Office of Special Investigations agents working out of Kirtland Air Force Base. These airmen, who included the intelligence agents who had first recruited Bill Moore to spy on Paul Bennewitz, were, by all accounts, imaginative. They had fabricated convincing-looking classified AFOSI reports that "revealed" sightings and bizarre UFO research projects; they had spread fantastic stories about cattle surgically massacred by alien invaders; they had even brought carefully targeted UFO believers to meet with them in restricted areas on government bases. The FBI had little doubt these AFOSI agents were capable of putting together the MJ-12 documents. But still unresolved was—why? Here, the FBI's inquiry began to falter. Perhaps, the Bureau suggested, the MJ-12 documents were another inspired training exercise, more on-the-job experience for fledgling Air Force agents in spreading disinformation. Or, another popular school of thought, these particular Air Force agents were a mean crowd; they were having a ball tormenting the UFO faithful. The FBI, however, was having a difficult time making any kind of case against the AFOSI officers. Every one of them, to a man, firmly denied he had a role in writing the MJ-12 documents. And, to complicate matters further, many of the agents had suddenly decided to retire. They now were private citizens, with, they emphatically emphasized (and on at least one occasion a lawyer was brought in to reiterate), all the constitutional rights of private citizens.

Which left, Colonel Phillips revealed, the FBI convinced that only one other source might have been responsible for the MJ-12 documents—the UFO Working Group.

There was some laughter from around the table; even the colonel allowed himself a smile. But before things got out of control, he waved an admonishing hand and insisted that the Working Group look at the possibility from the Bureau's perspective. Here was a group of high-level intelligence operatives—of course they had the talent to whip up a set of first-class phony government documents. They had a motive, too. If the UFO community as well as the press could be kept busy searching for proof about a mythical MJ-12 group, then a genuine Working Group would certainly have a better shot at continuing its own covert UFO investigation without detection or harassment. The colonel insisted that, on paper at least, it made sense; and heads nodded in agreement. But he added, clowning it up a bit with a heaving shrug of his shoulders and a bemused grin, "Only it just didn't happen that way. Honest, we didn't do it."

Once again there was a flutter of laughter. The idea that the FBI seriously considered the men in the Tank as suspects was greeted with great amusement. Colonel Phillips, playing to this mood, suggested that with the Working Group's "humble consent" he would try to persuade the FBI to concentrate its investigative energies on either of the two alternative scenarios.

That was when a DIA scientist sitting directly across the table spoke up. The colonel, he said, had started him thinking. To his mind there was a fourth possible scenario: The UFO Working Group itself might be the pretense. Its existence could be part of the cover-up.

The idea seemed to take everyone, including the speaker, by surprise. He waited a moment, as though he was unsure whether to continue, but in the end he couldn't contain himself. He went on with some emotion. Perhaps our job is to provide realistic cover, to keep the press, the UFO believers, and even the FBI and other intelligence agencies away from the real story. Perhaps our role is to go skulking off to the Mojave Desert or to Cheyenne Mountain or to Elmwood, Wisconsin, and quietly ask our questions. After all, if we don't know, who would? Yet all the while, somebody already knows. There's an MJ-12 Group or a whatever group somewhere in some dark corner of the chain of command that already knows exactly what is out there. They know that UFOs have visited this planet, that crashed alien spaceships have been recovered. And the very fact that we exist, that we meet in the Pentagon, that we go about our business so determined to get to the bottom of things, helps to protect their secret—UFOs exist.

"Impossible," Colonel Phillips shouted angrily. "I would have to know about it."

But this time no one laughed.

It was a time of growing suspicions. When I asked an FBI spokesman about the status of the MJ-12 inquiry, he turned brusque. Yet he promised he would get back to me within a day. He did. "Either on or off the record," he lectured, "the Bureau has no official interest" in the documents. When I began to argue that I knew that was not true, he became more emphatic. "Neither Washington nor any of the Bureau field offices are involved in any investigation" into the authenticity of the MJ-12 papers.

The next day an agent assigned to the Foreign Counterintelligence squad in New York spent a free-and-easy ten minutes talking to me about his active role in the MJ-12 case. Abruptly, however, his narrative came to a halt; perhaps I betrayed too much excitement. The press officer, he decided, his voice now filled with considerable anxiety, should handle all further questions. This official considered my request for four days before responding: "The alleged MJ-12 papers are part of an on-going classified investigation; therefore, the Bureau declines to comment."

Two months later, an exasperated FBI agent in Los Angeles shared more of the Bureau's dilemma: "We've gone knocking on every door in Washington with those MJ-12 papers. All we're finding out is that the government doesn't know what it knows. There are too many secret levels. You can't get a straight story. It wouldn't surprise me if we never know if the papers are genuine or not."

Yes, I glumly thought, I have been there, too. I have had the same official doors slammed in my face. For an instant I considered sharing the broken course across America I had been forced to run, the half-truths and lies I had been fed about the government's search for extraterrestrial life. I thought of telling him, another searcher, of my growing belief that the truth was deliberately being withheld.

But I kept my silence. There was, after all my tense encounters with people on the inside, after the specious and mercurial "on-the-record statements" that had been delivered up by the Bureau, *his* agency, a part of me that was cautious. I did not trust him.

He seemed to understand my quiet. "I want to get to the bottom of this as much as you do," he insisted.

"I'm sure," I said evenly.

Moments later we parted without saying good-bye.

Part VI
COMMUNION

THIRTY-NINE.

It was a wet, cold Washington noon and I was standing, as instructed, in front of the restaurant. It would have made more sense to have found shelter in the warm, dry lobby; who could be more conspicuous, after all, than the apparent lunatic with dripping hair and waterlogged tweeds slouching nonchalantly against a lamppost as if oblivious to the sudden downpour? But, though caught umbrellaless, I was reluctant to leave my (lamp)post. I had my orders. A source, as always, is the star. So let it rain. I would wait, as told, "by the lamppost out front at the crack of noon."

A bevy of secretaries—high-voiced, perfumed, full of Friday payday giggles—scurried by, while I anxiously noted that noon had "cracked" and then some. Still, I did not budge. It had taken me too long to get to this rain-splattered corner on the fringes of Georgetown. It was nearly three long years ago in this same city that I had stumbled onto the existence of the UFO Working Group. And now—at last—this meeting was looming in my hopeful mind as the final stop on a long and winding trail across America.

Standing there in the teeming rain, I found myself retracing my itinerary in my mind. Each leg of my journey had been an attempt to follow in the faint footsteps of the Working Group, and what a trip it had been. First to Cheyenne Mountain, a snow-covered tomb more impenetrable than any sandy pyramid along the Nile, its honeycomb of secret chambers filled, too, with priceless treasures. Then, by special NASA jet to the Mojave Desert, a vast, lonely, almost prehistoric landscape, with gigantic radio antennas standing in its midst, the twentieth century's most sophisticated attempts to eavesdrop on the future. Next I headed across California to the Moffett Field Naval Base, where SETI scientist Jill Tartar's office was festooned with a banner that proclaimed I BELIEVE IN LITTLE GREEN WOMEN, yet all the alien models in the base's exobiology laboratory appeared definitely to be male.

And then I backtracked south again to the Jet Propulsion Laboratory in Pasadena and its Digital Projects Lab, where two white-coated engineers exhibited the Multi channel Spectrum Analyzer with the pride other men might reserve for a much accomplished and beloved offspring, which I realized was not far from the actual case. Then off to little Elmwood with its whitewashed churches and rolling fields, where the villagers, direct and unembarrassed, chattered away with unrestrained certainty about hovering UFOs—but why anyone would choose to live in New York City: that, they confessed, was the truly unfathomable mystery. And not long after that I was sitting in a Naugahyde-covered booth in an Italian restaurant not a five-minute drive from the movie studio that had helped give birth to *Close Encounters of the Third Kind*, where I listened to a dour Bill Moore and an energetic Jaime Shandera detail an even more fantastic story—yet one they insisted was true. And from there I went back across the country to Harvard, where the inventor of Suitcase SETI recalled with genuine delight the time Leonard Nimoy, *Star Trek*'s Mr. Spock, had come to visit his lab. Then to MIT, where I caught the visionary Phillip Morrison typing away on his old Remington because his desktop computer was too cantankerous. Only to return again to the West, and a New Mexico night when the sky was so lovely, dark, and deep, and the Moon so perfect a golden ball that life on Earth seemed incomplete without the fulfilling promise contained in the heavens. And I continued on. There had been other destinations, other people eager to help.

But each time as I, notebook in hand, moved through the corridors of power, listened to scientists lecture, sat in farmhouses and offices asking my questions, I could not, no matter how inventively I tried, avoid slamming into sudden dead ends. The whole story was always lingering, deliberately, I came to believe, just out of my grasp.

Why?

This was the single, practical, impossible question that was balanced ominously on the tall peak of all my mounting suspicions. Why were all these official spokesmen and institutions doing their collusive best to hinder and obstruct my efforts? Why were stories true one day, and false the next? Why all the tense, unyielding secretiveness? Why were military intelligence agents spreading disinformation, driving UFO believers mad? What had the government found out there? What was it trying to hide?

And standing in the noontime rain, I was, I wanted to believe, on the verge of learning the answer. Earlier in the week I had been put in (the proverbial) touch with an individual who had lived an active life in intelligence circles; his résumé included, among more covert assignments, service with the CIA as well as a staff assignment to the National Security Council. When, after the usual tug-of-war, I had managed to steer the conversation around to the UFO Working Group, he conceded that he was familiar with its activities. "Could we meet?" I quickly asked. He, after some thought, agreed and, after even more deliberation, chose the time and place.

So I now waited by the lamppost. A good twenty minutes after the specified "crack of noon," a jaunty little figure in a Burberry and carrying an open red umbrella sauntered up to me and, full of apparent concern, smoothly exclaimed, "Howard? Glad to meet you. Now get underneath this before you catch your death." And I silently noted he had approached from the restaurant's lobby, where no doubt he had been waiting and watching all along.

We did not dine at the restaurant by the lamppost. "Let's find a place a bit more off the beaten path," he suggested with a genial smile. We headed, as he had clearly intended from the start, to a nearby Mexican restaurant. While we walked, he outlined a small cover story in the event of a chance encounter with anyone either of us knew. He was selling his house in Alexandria and I was a potential buyer. I readily agreed. In fact, all these elaborate precautions only served to increase my expectations of what was ultimately to be delivered. Who, after all, would value the secret that was divulged without the slightest struggle?

So I played along. He ordered a margarita, I ordered a margarita. He made small talk, I tried to carry my end. Truth was, he was fun to talk to; his conversation mixed a facile wit with a more combative sense of the ironic. He was very cynical about official Washington. And very funny. Sometimes after he finished telling a small story, he would raise his gray eyebrows as if he had been surprised by his own words. When he laughed, which was often, I laughed along with him. And I waited.

It wasn't until the meal was nearly over, while sipping coffee laced sweetly with Kahlua, that he, his voice soft but firm, asked that I tell him what I knew about the UFO Working Group.

I told him some.

When I had finished, he offered, "So you really have been busy." I took that as praise. Then he added, his eyebrows shooting upward, "Is that really all you've found out?"

I thought about his question. To receive, you sometimes have to give. Besides, I had nothing to lose. So I shared some of the obstacles I had encountered, the reasons for my suspicions.

All at once he was animated. His rigid control broke. "I knew it!" he nearly shouted, startling me in the process. "I knew the government was holding back. They know. They're just not telling us. Your story proves it."

But not to me. I asked for the check.

And so one more lead sank into ridiculousness. One more promising encounter ended up in farce. If this were an ancient parable, one of those medieval tales about a journey in search of a higher spiritual life, the traveler, after receiving help from wise men and magicians, after slaying dragons and false prophets, would at the end of his long and arduous adventures be magnificently rewarded—enlightenment would be his. This was not that kind of drama. We live in much more complicated times. Besides, this is, most inconveniently, a true story.

I had worked my way through dozens of official barriers, documented previously undisclosed levels of governmental interest and involvement in the search for extraterrestrial life, only to find that one more tormenting question still remained. This last, hard secret was impenetrable. Why was the government so committed, so conspiratorially determined, to perpetuate the mystery surrounding its fascination with other worlds? Why was the government not telling us everything? The more I puzzled over these questions, the more answers I had to grapple with.

Because keeping secrets is a habit. It is the way officials—spies, generals, and scientists—are taught to behave.

Because some explanations are not simple. All is never explained.

Because now that we are at the end of history, at the end of a politics of global conflict, as men and states abandon their allegiances to failed ideological gods, all that is left for a great nation to protect and believe in is its tattered secrets.

Because when we don't know, we assume that someone does. To paraphrase Chesterton, when people cease to believe their leaders, they come to believe not in nothing but in anything.

Because the glamour of a cover-up gives a daring and a coherence to our irrational fears. A cover-up is an easy answer.

Or, because they know. Because for our own safety, or in the national interest, or even for our own psychic good—they have decided not to tell us what is out there.

Any one of these answers might be true, I realized. Or some combination of them all. Or none. *Even the government doesn't know what it knows,* the FBI agent had warned me. How was I to have known those were words to live by? In a brief conversation with an NSA official nearly three years ago, I had blundered into the hardest story of them all. And perhaps the most despairing. For this was, I was now coming to believe, a tale that might have sprung forth from the dark imagination of a Beckett. Only now, the real-life actors do not simply wait impotently for Godot, but, resigned and helpless, they lament that this mysterious yet fulfilling stranger has indeed come—yet it doesn't matter. A malicious clique of the powerful has conspired to keep this miracle a secret.

But, in the end, I decided I was wrong. This was not a story about despair. For the more I thought about it, the more I understood that while working my way across America, while following the covert footsteps of the UFO Working Group, I had discovered something quite wonderful in the hearts and minds of those I encountered. It was more than a simple matter of shared belief, of a mutual instinct to turn their eyes and imaginations and hopes to the heavens because life on this planet was not enough—or, perhaps, had proved too much. It was more than a common philosophy about the supernal value and ubiquity of life. Those I interviewed shared more than a driving, tenacious curiosity, or an eager optimism. The tie went beyond psychology, beyond a preoccupation with the same fundamental symbols, archetypes, and psychic facts.

I had discovered a communion of spirit: a common will.

It was this conscious decision to act on their romantic ideals and daydreams about other worlds, this leap of real courage, that distinguished the people—the physicists, farmers, generals, exobiologists, UFO researchers, town fathers—I had met in my journey across America. They all were gifted with different abilities, different minds, different educations; nevertheless, it was this practical commitment to finding proof, to making contacts with other souls across the universe, that was their common bond, and the shared calling that allowed them to transcend bitter, perhaps even paranoid, despair. In dreams, they all agreed, began their responsibilities. And so, each in his own way, they went to work.

Yet, are the hopes in NASA's Project Cyclops report, a plan for a ten-billion-dollar dedicated SETI facility, really so different from those expressed in the spiral-bound booklet that urged the construction of a fifty-million-dollar UFO Landing Field?

Are the aspirations of the group of intelligence officers secretly meeting in the Pentagon vault room in conflict with those of Bill Moore's investigative team?

Is the Dolphin scientists' deduction that a message from another world will arrive only over the narrow range of "water hole" frequencies any less romantic a notion than the belief that an alien spacecraft will decide to land in Elmwood?

If a Harvard professor can build his own device to listen to space, are Paul Bennewitz's motives any less inspirational?

Were not Phillip Morrison's concert hall reverie and Tom Weber's nightmare sparked by the same unconscious desire to reach out to other worlds?

Are not the speculative Drake Equation and Stanton Friedman's exhaustive research papers both wishful attempts to make philosophy real?

Wasn't it the same humanistic goal, the hope that mankind might learn how to live with its ability to destroy itself, the *L* of the Drake Equation, that inspired both Congress to support SETI and Tom Weber to plan his landing field?

And, ultimately, are not the Drakes and the Morrisons, the Moores and the Shanderas, the Webers and the Colonel Phillipses, all adventurers, all explorers off on the same elusive quest?

But there is more. They share a common strength. They are steeled by more than simple courage. They have all experienced a large leap of the soul: a movement beyond resigned belief into an arena of absolute faith. They all know, in their secret hearts, that what they believe in, despite all odds, despite all obstacles of reason, will someday, no matter what, come to pass. We are not alone in the universe. They know, in time, there will be proof.

And so they are prepared to continue. To believe. And to wait.

This is a tale, then, where history is being made by men and women who dare to ask questions they know cannot yet be answered. It is a story by necessity without an ending. All its large figures are antiheroes; they have yet to win. They are victims of the same dilemma. They are confident they are right, but the proof eludes them. They are stuck at the same crossroads, waiting.

The UFO Working Group continues to serve President George Bush as it did President Reagan. Lately it has turned its attentions to the reports that a spacecraft landed in the Russian city of Voronezh. Colonel Phillips has doubts that the event actually happened, but still something in the Tass dispatch brought him up short. Genrikh Silanov, head of the Voronezh Geophysical Laboratory, revealed that the landing site was identified "by means of bilocation." Do the Russians have their own Project Aquarius? Phillips wondered to the Working Group. Are they also using scannate or some form of remote viewing to locate flying saucers? He decided that the Working Group should investigate.

Meanwhile in Elmwood, the success of UFO Days '88 convinced a lot of people that change was coming to the Eau Galle River Valley. Over at Thompson's Oil and Gas, an old pump was painted a bright red and, after a boisterous ceremony, hoisted up on the roof. "UFO Fuel," a sign written in slanted black letters archly explained. On the other side of Main Street, down at the Shack, Sissy rigged up a silver flying saucer on the roof. When she gave the word, a switch was thrown: the "UFO" actually spun in a lopsided circle. And Mayor Feiler, without even consulting the council, commissioned a local house painter "to design and erect suitable monuments" to commemorate the most notable sightings in the township's history. The painter was industrious and in no time at all four-foot-wide plywood models of dome-shaped spacecraft began appearing. Out by the quarry, for example, a metallic-painted plywood craft was nailed to two fence posts. Across the ship, in precise block letters, were the words:

UFO Sighting
April 22, 1976
George Wheeler
Police Officer

But, many people in Elmwood felt, the prospect of Tom Weber's UFO Landing Field promised something more momentous for their little town—prosperity. "Everywhere you went, people were starting to count money they hadn't made," said Mayor Feiler. "You'd go into any of the taverns and the first thing people would ask me, 'What's this I hear about the Holiday Inn checking out a site over on Shaw?' Or did I know anything about some plan to build a mall right by the banks of the Eau Galle. I mean, a *mall* in Elmwood, for God sakes. There'd be more stores than people."

"Not a day would go by," the mayor continued with some amusement, "when some farmer wouldn't call me and ask, 'Hey, Larry, you know if that Tom Weber got his land yet? 'Cause if he hasn't, I just happen to be sitting on a few spare hundred acres or so that would make a jim-dandy landing field for these here spacemen.' "

Tom Weber, though, had found his site. The acreage was just a quick ride from Main Street, but the route took you up a steep, twisting road lined with tall pines so that when you finally came to the top of the hill and gazed out on the rolling meadows, it was like seeing a brightly lit stage at the end of a long dark auditorium. The drama of the setting had convinced him. He was certain: *They* would choose to come to this soybean field a couple of hundred yards from Rock Elk Road, in the township of Elmwood, in the United States of America, on the planet Earth. This was where he would build his landing field.

The farmer who owned the 160 acres was willing to sell—at a price that was, at least the way Tom Weber appraised it, more than three times the going rate for tillable land. "The Site Center's got a long way to go before it'll ever be able to buy land for that kind of money" was his only response. And later he added, "Greed—that's what's going to destroy this town. I'm offering Elmwood the future, but all people here see is a chance to make a buck."

Still, he has not given up hope. Perhaps someone—he still hopes for a Steven Spielberg, a Donald Trump—will come to his rescue. How could they not? How could they ignore what is at stake? In the meantime, until he can actually own the land and begin to build, he is content to visit the site. He goes at night, most often by himself. He sits cross-legged in a hilltop field, alone, surrounded by total quiet, the smell of soybeans still strong even though harvest was a warm month ago. His neck is bent, and he is looking up at the nighttime sky. He is watching, searching, and that, he finds, is very comforting.

Bill Moore's team, too, is searching. All of Phil Klass's white papers disputing the authenticity of the MJ-12 documents, they are convinced, have proved nothing. It will only be a matter of time before the truth is known. Moore says his work on the inside, his meetings with the Falcon, have left him convinced that the government has already been contacted by extraterrestrials. The secret cannot be held forever. "Someone in the government will be coming forward shortly," Shandera promises.

And Kent Cullers, the heir to Morrison's, Cocconi's, and Drake's vision: How does he spend his share of the rich legacy left by the Order of the Dolphin? He listens, and he, too, dreams. Yet, is he simply—as some debunker might cruelly sneer—just one more blind man looking for what he never will be able to see? No, I do not think so. For now, at the end of my journey, I have become a believer. There are other worlds. I am certain the day will come when Cullers or one of the other blind men will be sitting in the white trailer in the boiling Mojave and he will, for the first time in his life, see. A noise, life shaped into a beautiful music, will travel across ink-black space and time, and into legend.

And then, just as they all had expected, this story will end, and the future will finally begin.

Appendix
THE MJ-12 DOCUMENTS

TOP SECRET / MAJIC
EYES ONLY
NATIONAL SECURITY INFORMATION

****
* TOP SECRET *
****

EYES ONLY COPY <u>ONE</u> OF <u>ONE</u>.

BRIEFING DOCUMENT: OPERATION MAJESTIC 12

PREPARED FOR PRESIDENT-ELECT DWIGHT D. EISENHOWER: (EYES ONLY)

18 NOVEMBER, 1952

<u>WARNING</u>! This is a TOP SECRET - EYES ONLY document containing compartmentalized information essential to the national security of the United States. EYES ONLY ACCESS to the material herein is strictly limited to those possessing Majestic-12 clearance level. Reproduction in any form or the taking of written or mechanically transcribed notes is strictly forbidden.

****
* TOP SECRET *
****

TOP SECRET / MAJIC
EYES ONLY

EYES ONLY T52-EXEMPT (E)

TOP SECRET / MAJIC
EYES ONLY

****
* TOP SECRET *
****

EYES ONLY COPY ONE OF ONE.

SUBJECT: OPERATION MAJESTIC-12 PRELIMINARY BRIEFING FOR PRESIDENT-ELECT EISENHOWER.

DOCUMENT PREPARED 18 NOVEMBER, 1952.

BRIEFING OFFICER: ADM. ROSCOE H. HILLENKOETTER (MJ-1)

NOTE: This document has been prepared as a preliminary briefing only. It should be regarded as introductory to a full operations briefing intended to follow.

* * * * * *

OPERATION MAJESTIC-12 is a TOP SECRET Research and Development/ Intelligence operation responsible directly and only to the President of the United States. Operations of the project are carried out under control of the Majestic-12 (Majic-12) Group which was established by special classified executive order of President Truman on 24 September, 1947, upon recommendation by Dr. Vannevar Bush and Secretary James Forrestal. (See Attachment "A".) Members of the Majestic-12 Group were designated as follows:

Adm. Roscoe H. Hillenkoetter
Dr. Vannevar Bush
Secy. James V. Forrestal*
Gen. Nathan F. Twining
Gen. Hoyt S. Vandenberg
Dr. Detlev Bronk
Dr. Jerome Hunsaker
Mr. Sidney W. Souers
Mr. Gordon Gray
Dr. Donald Menzel
Gen. Robert M. Montague
Dr. Lloyd V. Berkner

The death of Secretary Forrestal on 22 May, 1949, created a vacancy which remained unfilled until 01 August, 1950, upon which date Gen. Walter B. Smith was designated as permanent replacement.

****
* TOP SECRET *
****

TOP SECRET / MAJIC
EYES ONLY

EYES ONLY T52-EXEMPT (E)

TOP SECRET / MAJIC
EYES ONLY

* TOP SECRET *
****

EYES ONLY   COPY ONE OF ONE.

On 24 June, 1947, a civilian pilot flying over the Cascade Mountains in the State of Washington observed nine flying disc-shaped aircraft traveling in formation at a high rate of speed. Although this was not the first known sighting of such objects, it was the first to gain widespread attention in the public media. Hundreds of reports of sightings of similar objects followed. Many of these came from highly credible military and civilian sources. These reports resulted in independent efforts by several different elements of the military to ascertain the nature and purpose of these objects in the interests of national defense. A number of witnesses were interviewed and there were several unsuccessful attempts to utilize aircraft in efforts to pursue reported discs in flight. Public reaction bordered on near hysteria at times.

In spite of these efforts, little of substance was learned about the objects until a local rancher reported that one had crashed in a remote region of New Mexico located approximately seventy-five miles northwest of Roswell Army Air Base (now Walker Field).

On 07 July, 1947, a secret operation was begun to assure recovery of the wreckage of this object for scientific study. During the course of this operation, aerial reconnaissance discovered that four small human-like beings had apparently ejected from the craft at some point before it exploded. These had fallen to earth about two miles east of the wreckage site. All four were dead and badly decomposed due to action by predators and exposure to the elements during the approximately one week time period which had elapsed before their discovery. A special scientific team took charge of removing these bodies for study. (See Attachment "C".) The wreckage of the craft was also removed to several different locations. (See Attachment "B".) Civilian and military witnesses in the area were debriefed, and news reporters were given the effective cover story that the object had been a misguided weather research balloon.

****
* TOP SECRET *
****

EYES ONLY TOP SECRET / MAJIC
EYES ONLY

T52-EXEMPT (E)

TOP SECRET / MAJIC
EYES ONLY

* TOP SECRET *

EYES ONLY

COPY ONE OF ONE.

A covert analytical effort organized by Gen. Twining and Dr. Bush acting on the direct orders of the President, resulted in a preliminary concensus (19 September, 1947) that the disc was most likely a short range reconnaissance craft. This conclusion was based for the most part on the craft's size and the apparent lack of any identifiable provisioning. (See Attachment "D".) A similar analysis of the four dead occupants was arranged by Dr. Bronk. It was the tentative conclusion of this group (30 November, 1947) that although these creatures are human-like in appearance, the biological and evolutionary processes responsible for their development has apparently been quite different from those observed or postulated in homo-sapiens. Dr. Bronk's team has suggested the term "Extra-terrestrial Biological Entities", or "EBEs", be adopted as the standard term of reference for these creatures until such time as a more definitive designation can be agreed upon.

Since it is virtually certain that these craft do not originate in any country on earth, considerable speculation has centered around what their point of origin might be and how they get here. Mars was and remains a possibility, although some scientists, most notably Dr. Menzel, consider it more likely that we are dealing with beings from another solar system entirely.

Numerous examples of what appear to be a form of writing were found in the wreckage. Efforts to decipher these have remained largely unsuccessful. (See Attachment "E".) Equally unsuccessful have been efforts to determine the method of propulsion or the nature or method of transmission of the power source involved. Research along these lines has been complicated by the complete absence of identifiable wings, propellers, jets, or other conventional methods of propulsion and guidance, as well as a total lack of metallic wiring, vacuum tubes, or similar recognizable electronic components. (See Attachment "F".) It is assumed that the propulsion unit was completely destroyed by the explosion which caused the crash.

* TOP SECRET *

EYES ONLY TOP SECRET / MAJIC T52-EXEMPT (E)

EYES ONLY

TOP SECRET / MAJIC
EYES ONLY

* TOP SECRET *

EYES ONLY

COPY ONE OF ONE.

A need for as much additional information as possible about these craft, their performance characteristics and their purpose led to the undertaking known as U.S. Air Force Project SIGN in December, 1947. In order to preserve security, liason between SIGN and Majestic-12 was limited to two individuals within the Intelligence Division of Air Materiel Command whose role was to pass along certain types of information through channels. SIGN evolved into Project GRUDGE in December, 1948. The operation is currently being conducted under the code name BLUE BOOK, with liason maintained through the Air Force officer who is head of the project.

On 06 December, 1950, a second object, probably of similar origin, impacted the earth at high speed in the El Indio - Guerrero area of the Texas - Mexican boder after following a long trajeotory through the atmosphere. By the time a search team arrived, what remained of the object had been almost totally incinerated. Such material as could be recovered was transported to the A.E.C. facility at Sandia, New Mexico, for study.

Implications for the National Security are of continuing importance in that the motives and ultimate intentions of these visitors remain completely unknown. In addition, a significant upsurge in the surveillance activity of these craft beginning in May and continuing through the autumn of this year has caused considerable concern that new developments may be imminent. It is for these reasons, as well as the obvious international and technological considerations and the ultimate need to avoid a public panic at all costs, that the Majestic-12 Group remains of the unanimous opinion that imposition of the strictest security precautions should continue without interruption into the new administration. At the same time, contingency plan MJ-1949-04P/78 (Top Secret - Eyes Only) should be held in continued readiness should the need to make a public announcement present itself. (See Attachment "G".)

* TOP SECRET *

TOP SECRET / MAJIC
EYES ONLY

EYES ONLY

T52-EXEMPT (E)

TOP SECRET / MAJIC
EYES ONLY

* TOP SECRET *

<u>EYES ONLY</u>　　　　COPY <u>ONE</u> OF <u>ONE</u>.

ENUMERATION OF ATTACHMENTS:

*ATTACHMENT "A"........Special Classified Executive Order #092447. (TS/EO)

*ATTACHMENT "B"........Operation Majestic-12 Status Report #1, Part A. 30 NOV '47. (TS-MAJIC/EO)

*ATTACHMENT "C"........Operation Majestic-12 Status Report #1, Part B. 30 NOV '47. (TS-MAJIC/EO)

*ATTACHMENT "D"........Operation Majestic-12 Preliminary Analytical Report. 19 SEP '47. (TS-MAJIC/EO)

*ATTACHMENT "E"........Operation Majestic-12 Blue Team Report #5. 30 JUN '52. (TS-MAJIC/EO)

*ATTACHMENT "F"........Operation Majestic-12 Status Report #2. 31 JAN '48. (TS-MAJIC/EO)

*ATTACHMENT "G"........Operation Majestic-12 Contingency Plan MJ-1949-04P/78: 31 JAN '49. (TS-MAJIC/EO)

*ATTACHMENT "H"........Operation Majestic-12, Maps and Photographs Folio (Extractions). (TS-MAJIC/EO)

* TOP SECRET *

TOP SECRET / MAJIC
EYES ONLY

<u>EYES ONLY</u>　　　　T52-EXEMPT (E)

TOP SECRET / MAJIC
EYES ONLY

* TOP SECRET *

EYES ONLY COPY **ONE** OF **ONE**.

ATTACHMENT "A"

* TOP SECRET *

EYES ONLY TOP SECRET / MAJIC
EYES ONLY

T52-EXEMPT (E)

TOP SECRET
EYES ONLY

THE WHITE HOUSE
WASHINGTON

September 24, 1947.

MEMORANDUM FOR THE SECRETARY OF DEFENSE

Dear Secretary Forrestal:

As per our recent conversation on this matter, you are hereby authorized to proceed with all due speed and caution upon your undertaking. Hereafter this matter shall be referred to only as Operation Majestic Twelve.

It continues to be my feeling that any future considerations relative to the ultimate disposition of this matter should rest solely with the Office of the President following appropriate discussions with yourself, Dr. Bush and the Director of Central Intelligence.

Harry Truman

TOP SECRET
EYES ONLY

July 14, 1954

~~TOP SECRET RESTRICTED~~
~~SECURITY INFORMATION~~

MEMORANDUM FOR GENERAL TWINING

SUBJECT: NSC/MJ-12 Special Studies Project

The President has decided that the MJ-12 SSP briefing should take place during the already scheduled White House meeting of July 16, rather than following it as previously intended. More precise arrangements will be explained to you upon arrival. Please alter your plans accordingly.

Your concurrence in the above change of arrangements is assumed.

ROBERT CUTLER
Special Assistant
to the President

List of Abbreviations and Acronyms

AFB	Air Force Base
AFIS	Air Force Intelligence Service
AFOSI	Air Force Office of Special Investigations
APRO	Aerial Phenomena Research Organization
ASAT	Anti-Satellite Weapon
ASW	Anti-Submarine Warfare
CIA	Central Intelligence Agency
CINC	Commander in Chief
cm	centimeter
COMIREX	Committee on Imagery Requirements and Exploitation
DCD	Domestic Collection Division (CIA)
DIA	Defense Intelligence Agency
DOD	Department of Defense
ELINT	Electronic Intelligence
ESI	Extremely Sensitive Information
exobio	exobiology
FBI	Federal Bureau of Investigation
FCI	Foreign Counterintelligence (FBI)
FOIA	Freedom of Information Act
FTD	Foreign Technology Division (Air Force)
GHz	gigahertz
HUMINT	Human Intelligence
INSCOM	Intelligence and Security Command (Army)
JCS	Joint Chiefs of Staff
JPL	Jet Propulsion Laboratory
KH	Keyhole Satellite
KHz	kilohertz
M and O	Directorate for Management and Operations (DIA)
MHz	megahertz

MIRV	multiple independently targeted reentry vehicles
MJ-12	Majestic-12 documents
NASA	National Aeronautics and Space Administration
NICAP	National Investigations Committee on Aerial Phenomena
NMIC	National Military Intelligence Center
NORAD	North American Air Defense Command
NRO	National Reconnaissance Office
NSA	National Security Agency
NSC	National Security Council
OSI	Office of Strategic Information
OSS	Office of Strategic Services
OST	Office of Science and Technology (President's)
PDB	President's Daily Brief
RPV	Remotely Piloted Vehicle
RV	Remote Viewing
SAC	Strategic Air Command
SCI	Sensitive Compartmented Information
SETI	Search for Extraterrestrial Intelligence (NASA)
SIGNIT	Signals Intelligence
SPADATS	Space Tracking and Detection System
SRI	Stanford Research Institute
UFO	Unidentified Flying Object

A Note on Sources

A reporter's search for truth—that is, what *actually* happened—is always a rough journey, but it becomes a particularly rocky trip when the participants in the story are doing their resolute best to hold on to their secrets. That was, as I have detailed in these pages, the formidable obstacle I had to contend with as I tried to discover the history of the Defense Intelligence Agency's UFO Working Group and its investigations into the possibility of extraterrestrial life. In fact, so pervasive, so determined, was the commitment to secrecy and obfuscation that I grew to realize this perplexing official mind-set was a valid—even essential—part of my story. If I was to tell this story accurately, I had no choice but to enter the drama, to recount my role as the reportorial "I" who went knocking on every door, who was lied to, who wound up on a sort of Gulliver's travels to strange "lands" across America.

Yet once this narrative decision was made, it created a second problem. Would the tale become too subjective? Would it, stripped of the anonymous yet authoritative perspective of the familiar journalistic third person, read simply as one reporter's *version* of what happened? Or, worse, would the account seem too novelistic? Would readers understand that seemingly casual details—the shopping spree that delayed Commander Mondran (Chapter One), the Michelob key ring dangling from Bill Moore's jeans (Chapter Thirty-two), or even the greasy french fry waved threateningly across a table at a young scientist (Chapter Thirteen)—were rock-solid facts exacavated from mountains of research and observation and not simply an unscrupulous author's easy inventions?

Also, now is the time to admit there was another ambition shaping and complicating my telling of this story. The book attempts to show how the people—scientists, generals, farmers—who populate its pages think: how they construe the world and infuse their lives with meaning.

And to accomplish this, the book's style and structure were influenced by the narrative pace of popular science fiction—as were the real-life actions and beliefs of many of its heroes, both thinkers and dreamers.

Yet, this is a true story.

The reader, however, has a right to know what is meant by "true." The reader has a right to know if any liberties were taken in the pursuit of style, in the dramatized telling of the tale.

Here, then, are the standards I adhered to in writing the book. It is based largely on interviews. Over the nearly three years I spent chasing after the story, I interviewed 212 individuals. There were lengthy conversations, such as the one with Professor Paul Horowitz at Harvard that continued throughout two fascinating days; there were interviews that were conducted over a number of months, such as those with Mayor Larry Feiler of Elmwood, with whom I had three phone conversations in addition to spending an informative afternoon in his office; and there were numerous brief but helpful telephone interviews, such as my frequent calls to the press officers of the U.S. Space Command ("Box Nine is on what floor, please?"). Also, I have had no choice but to make use of sources who requested—often demanded—anonymity. This is, I will agree, one of the crutches of journalism and part of what separates the genre from the more rigidly standardized discipline of history. Yet, a reporter's covenant is to protect the privacy and livelihoods of those who have, often at some personal risk, befriended him with information. And his obligation is to tell his story while it is still fresh, while it is still news. That was my dilemma, and the book is a product of the choices I made.

Still, in using the material gathered in these interviews I have followed strict rules. If dialogue is quoted directly, the source is either the individual who is speaking, or someone who witnessed the conversation. And often I felt free to use the absolute authority of direct quotations because I had a variety of concurring sources—the speaker and two or more eyewitnesses. Other times, however, when my sources disagreed about what was said or could not recall the conversation precisely, my method was more judgmental. I tried to keep as close as possible to the general way my informants remembered events, and ultimately I favored reconstructions that in my opinion made the most sense. Direct quotation marks were never used in these situations. And when individuals disagreed over important points—for example, General Stubblebine's memory of any role he might have played with the 1983 UFO taskforce—I felt duty-bound to include both sides of these arguments.

Another important source of information was found in the tall piles of newspaper articles, magazine pieces, and government publications (both scientific documents and military "backgrounders") that I assembled in my years of research. These written sources were used to substantiate information collected in interviews (for example: Phil Klass's exhaustive articles in *The Skeptical Inquirer*; Stanton Friedman's privately published essays on "the cosmic Watergate," as well as those he published in *International UFO Reporter* on the MJ-12 evidence and "the secret life" of Donald Menzel; and monographs by Russell Targ and Dr. Harold Puthoff on Remote Vision). They also provided additional details as I fleshed out scenes (for example: privately published pamphlets on the history of Elmwood, Wisconsin; U.S. Navy "fact sheets" on the "electronic fence" detection system) and were essential to an understanding of the fundamentals of the science that has convinced NASA there is other life in the universe—and a "rational" way to search for it (Part III, The Order of the Dolphin).

My library of scientific source material grew to more than 200 books and articles, and it was an enjoyable education. If any reader is interested in taking a similar course, I would suggest he could not fail to be intrigued if his preliminary reading list included Walter Sullivan's *We Are Not Alone* (Signet Books), John Kraus's *Big Ear* (Cygnus-Quasar Books), NASA's *SETI* (government publication no. SP-419), and *Life in the Universe*, edited by John Billingham (NASA Conference Publication 2156). These books were supplemented by the several volumes of the now unfortunately defunct magazine *Cosmic Search*, which contained wonderfully candid reminiscences by Professors Frank Drake and Phillip Morrison that helped to shape my reporting in Part III.

Also, the description of Carl Sagan's confrontation with Senator Proxmire in Part III came largely from a speech the astronomer made in 1985 to the International Astronomical Union, as reported by Thomas R. McDonough in his witty *The Search For Extraterrestrial Intelligence* (John Wiley & Sons, Inc., 1987).

Additionally, it should be noted that Tom Weber's interrogator in Chapter Twenty-three was Marcia Nelesen of the *Janesville Sunday Gazette*, where that interview, in slightly different form, first appeared. Similarly, while many of Bill Moore's comments and observations (Part V, Counterintelligence) were made in the course of interviews with me, another important source for directly quoted information was the transcript of a long presentation he made to a UFO conference in Las Vegas in July 1989 (available through William L. Moore Publications). Also, as I made my way through government archives armed with the Freedom of Information Act, I relied greatly on the pioneering research done by Lawrence Fawcett and Barry J. Greenwood for their book *Clear Intent* (Prentice-Hall). And, as I indicate in the text, I also was able to review and quote from government documents that still remain classified.

Above all, I have done my best throughout the pages of this book to include the dates when events occurred, the names of the specific individuals who played a role in the story, and to make my source attributions immediately apparent. If, at times, I have found it necessary to protect the identity of an individual, it was a compromise I have openly shared with the reader. It was the only way secrets could be revealed, and a true story told.

Acknowledgments

While writing a book is often a solitary, lonely enterprise, publishing one is always a communal experience. And, in my case at least, that was a blessing. Many people have been very generous with their help and advice. I was fortunate to have Simon and Schuster as a home. Dick Snyder and Charlie Hayward were on my side from the moment when I first approached them with my inchoate idea, and they continued to offer encouragement throughout the entire process. My editor, Fred Hills, held me firmly to the task. He was, as the occasion demanded, stern, critical, encouraging, wise—and, throughout it all, a friend. Burton Beals's special intelligence was a godsend; I relied on his wisdom (and kindness) greatly. Daphne Bien was also gracious with her help; she managed to smile no matter how often I wound up pestering her in the course of a single day. And, once the manuscript was completed, Jack McKeown and Lisa Kitei were no less indulgent in putting up with an anxious author; they always tried to help.

I was also able to take advantage of the advice, succor, and goodheartedness of my agents. Lynn Nesbit and Suzanne Gluck were involved with the book from the start and they helped me throughout. And along the way, I profited from the attentions of Ed Olson, Heather Schroeder, and Lew Grimes.

Other early sources of support and encouragement were David Wolper and Bernie Sofronski at Warner Brothers Television. They committed their talents to making my book into a miniseries while it was still an unfinished manuscript. Equally important, Richard DeLong Adams, the screenwriter who adapted my last book for television, became involved in the project and he, as in the past, was kind enough to offer perceptive comments about my manuscript as it evolved into a book.

Not that it was always fun. But when things got complicated, Dan

Wolf always offered wise counsel and unswerving friendship; Phil Werber got me to laugh; my mother told me not to worry; my sister, Marcy, and her husband (and my friend), Peter Ashkenasy, listened to my woes and actually convinced me things weren't so bad; and Jenny filled me with hope and love.

About the Author

Howard Blum, a former award-winning reporter for *The New York Times*, has written several highly acclaimed and best-selling nonfiction books that not only are headline-making works of investigative journalism, but also tell compelling true stories. In *Wanted! The Search for Nazis in America*, Mr. Blum told a real-life detective story—and wrote a book that led to congressional hearings. In *I Pledge Allegiance . . . The True Story of the Walkers: An American Spy Family*, he wrote an espionage thriller about the most damaging spy ring in America's history. The book was the basis for the CBS miniseries, and Mr. Blum served as the drama's associate producer. And now in *Out There*, he reveals a true story about the future. Mr. Blum lives and works in New York, where he is presently involved in writing another true story.